Essential Biology

Eukaryotic Cell
&
Cellular Metabolism

3rd edition

STERLING
Education

3   2   1

ISBN-13: 979-8-8855717-3-9

Sterling Education materials are available at quantity discounts.

Contact info@sterling–prep.com

Sterling Education
6 Liberty Square #11
Boston, MA 02109

Published by Sterling Education

 Printed in the U.S.A.

STERLING
Education

From the foundations of a living cell to the complex mechanisms of gene expression, *Essential Biology Self-Teaching Guides* are a comprehensive compendium of clearly explained texts to learn and master multifaceted biology topics.

These guides provide a detailed review of essential biological processes of living systems. Develop a better understanding of cell and molecular biology, mechanisms of cell metabolism, plants and photosynthesis, evolution and natural selection, ecology and population biology. Learn the principles of genetics, microbiology, classification and diversity, as well as structure and function of anatomical systems. Reinforce your learning by working through the practice questions and detailed explanations.

Created by highly qualified biology instructors, researchers, and education specialists, these books empower readers by helping them increase their understanding of biology.

We sincerely hope that these guides are valuable for your learning.

230731akp

Featured on

Essential Biology Self-Teaching Guides

Eukaryotic Cell & Cellular Metabolism

Molecular Biology & Genetics

Nervous & Endocrine Systems

Circulatory, Respiratory & Immune Systems

Digestive & Excretory Systems

Muscle, Skeletal & Integumentary Systems

Reproduction & Development

Microbiology

Plants & Photosynthesis

Evolution, Classification & Diversity

Ecology & Population Biology

Visit our Amazon store

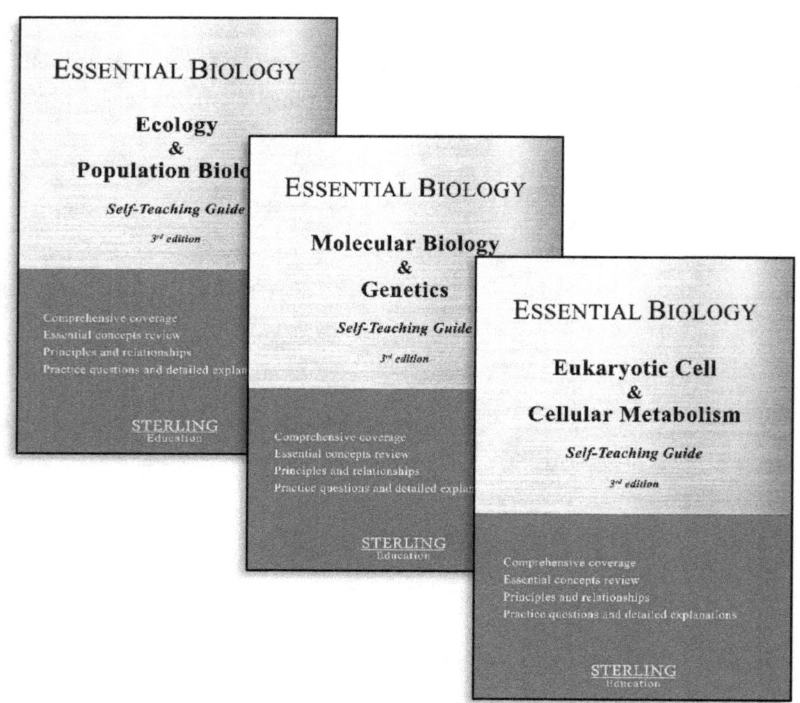

Essential Chemistry Self-Teaching Guides

Electronic Structure & Periodic Table

Chemical Bonding

States of Matter & Phase Equilibria

Stoichiometry

Solution Chemistry

Chemical Kinetics & Equilibrium

Acids & Bases

Chemical Thermodynamics

Electrochemistry

Visit our Amazon store

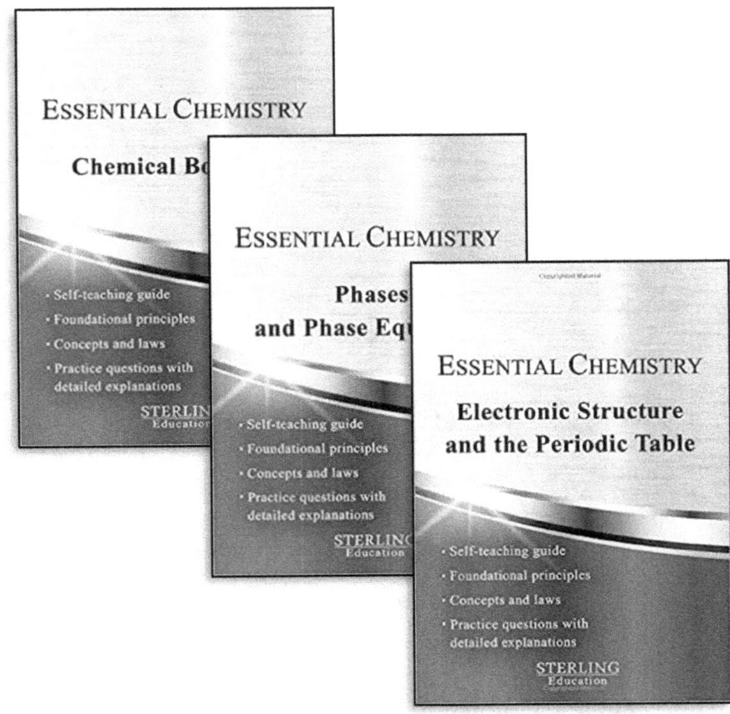

Kinematics and Dynamics

Equilibrium and Momentum

Force, Motion, Gravitation

Work and Energy

Fluids and Solids

Waves and Periodic Motion

Light and Optics

Sound

Electrostatics and Electromagnetism

Electric Circuits

Heat and Thermodynamics

Atomic and Nuclear Structure

Visit our Amazon store

Chemistry

Physics

Cell and Molecular Biology

Organismal Biology

American History

American Law

American Government and Politics

Comparative Government and Politics

World History

European History

Psychology

Environmental Science

Human Geography

Visit our Amazon store

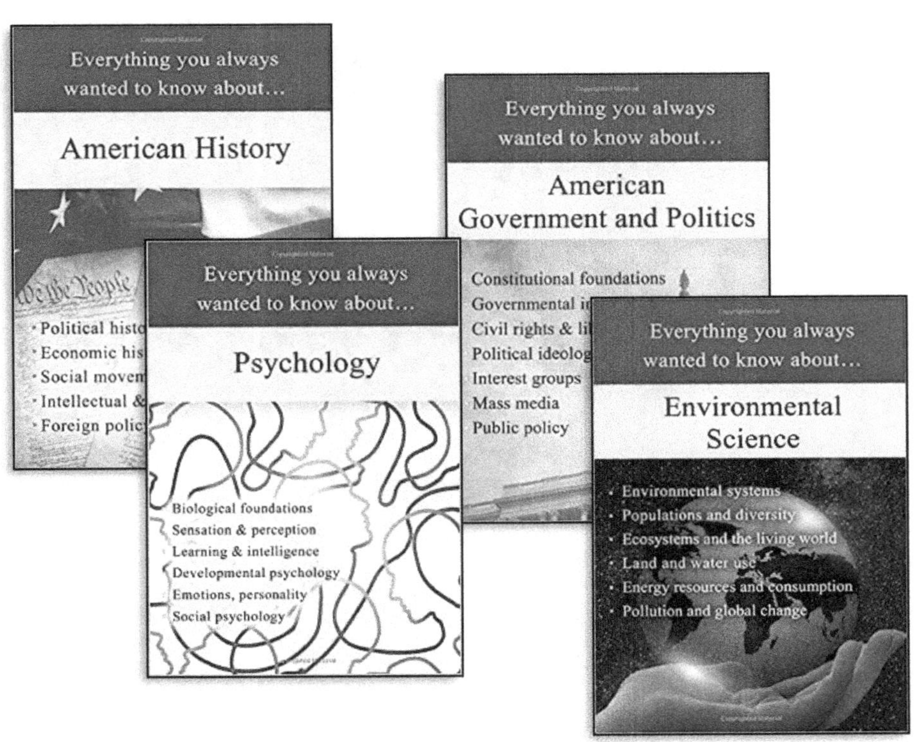

Page intentionally left blank

Table of Contents

Table of Contents (*continued*)

REVIEW: Eukaryotic Cell: Structure & Function (*continued*)

Table of Contents (*continued*)

Table of Contents (*continued*)

Table of Contents (*continued*)

Page intentionally left blank

REVIEW

Eukaryotic Cell:
Structure & Function

Cell Theory

Prokaryotic and Eukaryotic Cells

Membrane-Bound Organelles

Plasma Membrane

Cytoskeleton

Cell Cycle and Mitosis

Cell Cycle Control

Page intentionally left blank

Cell Theory

Defining characteristics of cells

Cell theory, the scientific theory which describes the morphological and biochemical properties of cells, is a fundamental doctrine of biology.

Classical cell theory includes three fundamental tenets derived from the research of early biologists:

1. *All living organisms are composed of one or more cells.*

    Multicellular organisms are composed of many cells.

    Unicellular organisms (e.g., bacteria) are composed of one cell.

2. *Cells are the smallest, basic units of life.*

    Cells are the smallest units of life because they are the smallest structures capable of carrying out the fundamental metabolic process (e.g., reproduce and divide, extract energy from their environment).

3. *Cells arise from pre-existing cells and cannot be created from non-living material.*

    Creating new cells is cellular division and used in sexual and asexual reproduction.

As the modern understanding of biology evolved, so did the tenets of cell theory.

Modern cell theory adds the following concepts to classical cell theory:

4. *Cells pass on the genetic material during replication in the form of DNA.*

5. *The cells of organisms are chemically similar.*

6. *Cells are responsible for energy flow and metabolism.*

The tenets of cell theory are somewhat dynamic, and as such, some scientists may omit some tenets or include others not mentioned here.

History and development

Robert Hooke first observed the small structural units, which he called "cells," under the light microscope in 1665. However, it was not until nearly 200 years later that significant progress was made in understanding these cells. In the 1830s, botanist Matthias Schleiden discovered cells in the tissues of plants and declared that cells are the building blocks of plants. Simultaneously, Theodor Schwann published his work on cell theory, generalizing it to plants and animals. Shortly after that, Rudolf Virchow overturned the predominating belief that cells were spontaneously generated from non-living matter. He proclaimed that all cells must arise from other living cells.

Cell biology was consistently advanced by new findings in membrane physiology, mitosis, and other cellular and molecular processes.

In the 1950s, James Watson and Francis Crick published their discovery of the molecular structure of DNA, which revolutionized the entire field of biology. Other researchers, including Rosalind Franklin, went uncredited for their seminal work on this monumental discovery.

Impact on biology

Cell theory is an important unifying concept that provides biologists with a common understanding from which they can make further discoveries. Once the foundation of cell theory was laid, along with advances in laboratory techniques and instrumentation, scientific progress in the field of biology rapidly accelerated. The relatively slow progress of biology before the mid-1800s demonstrates the difficulties of advancing in a field where the framework is not understood.

By understanding the basics of a single cell, researchers could add context to life sciences. This is because cells represent the building blocks of all organisms. Multicellular organisms exhibit *emergent properties*, meaning the whole is greater than the sum of its parts. Cells form tissues, tissues form organs, organs form organ systems, and organ systems form multicellular organisms. For example, the individual cells in the lungs are not of much use by themselves, but when combined as a working unit, they create a highly sophisticated set of lungs essential for the organism's survival.

Prokaryotic and Eukaryotic Cells

Nucleus and other defining cellular characteristics

Cells are divided into two taxa: *Eukarya* and *Prokarya.*

Prokaryotes are divided into the domains of Bacteria and Archaea.

Prokaryotes and eukaryotes' salient difference is that prokaryotes lack a nucleus and membrane-bound organelles. A prokaryotic cell's main intracellular components are its single double-stranded circular DNA molecule, ribosomes, and cytoplasm.

Prokaryotic cells include a phospholipid plasma membrane and an outer peptidoglycan cell wall. They are usually smaller than eukaryotic cells with smaller ribosomes (30S and 50S subunits; 70S as assembled).

Eukaryotic cells have linear DNA enclosed in a membrane-bound nucleus and membrane-bound organelles.

Eukaryotic cells replicate via mitosis or meiosis, while prokaryotes replicate via *binary fission,* a form of asexual reproduction.

Many similarities exist between the two cell types.

Prokaryotes and eukaryotes contain cytoplasm, ribosomes, and DNA, and are unicellular or multicellular, although multicellular prokaryotes are rare.

Some eukaryotes are capable of asexual reproduction, albeit differently from prokaryotes.

Plants and fungi (eukaryotes) have cell walls like prokaryotes.

Comparing prokaryotes and eukaryotes

Prokaryotes	Eukaryotes
Domains: Bacteria and Archaea	Domain: Eukarya
Cell wall present in all prokaryotes	Cell wall in fungi, plants, and some protists
No nucleus, circular strand of dsDNA	Membrane-bound nucleus housing dsDNA
Ribosomes (subunits = 30S and 50S; 70S)	Ribosomes (subunits = 40S and 60S; 80S)
No membrane-bound organelles	Membrane-bound organelles

Biochemistry considerations of surface-volume constraints

The cell's metabolic activity describes the biochemical reactions within the cell.

Substances need to be taken into the cell to fuel these reactions, while the reactions' waste products need to be removed.

When the cell increases in size, so do its metabolic activity.

The cell's surface area is vital because it affects the rate particles can enter and exit, with a larger surface area resulting in a higher uptake and excretion rate.

The volume affects the rate at which biochemical materials are made (or consumed) within the cell; the chemical activity per unit of time.

As the volume (and associated chemical activity) of the cell increases, so does the surface area, but not to the same extent. When the cell gets bigger, its surface area-to-volume ratio gets smaller.

If the surface area-to-volume ratio decreases, substances cannot enter the cell fast enough to fuel reactions.

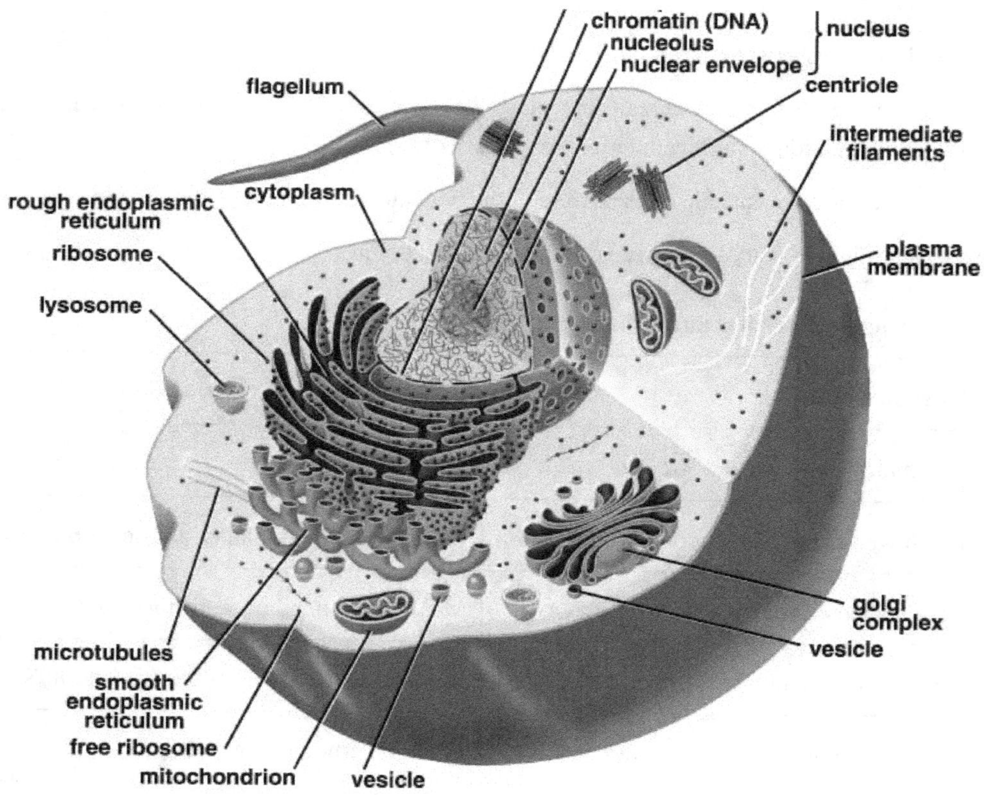

A eukaryotic animal cell with no cell wall or chloroplasts as in eukaryotic plant cells

Waste products are produced faster than they can be excreted, accumulating inside the cell. Cells are not able to lose heat fast enough and may overheat.

The surface-area-to-volume ratio is important for a cell. The physical limitation of the area-to-volume ratio limits the size of cells.

Nucleus compartmentalizes genetic information

The *nucleus* is the largest membrane-bound organelle in the center of most eukaryotic cells. It contains the cell's genetic code—its DNA. The nucleus's function is to direct the cell by storing and transmitting genetic information.

Cells contain multiple nuclei (e.g., skeletal muscle cells), one, or rarely, none (e.g., red blood cells).

Inside the nucleus is the *nuclear lamina,* a dense network of filamentous and membranous proteins associated with the nuclear envelope and its pores.

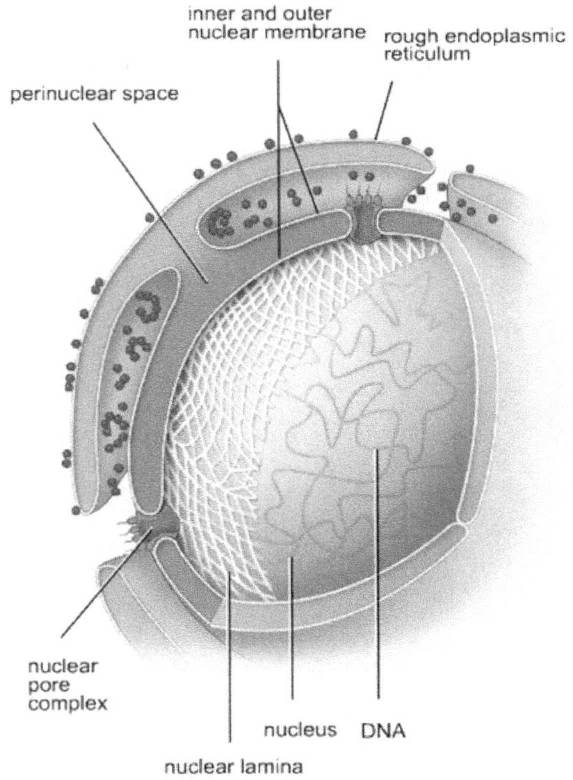

A double membrane surrounds the nucleus of the cell with nuclear pores
for select transport of a substance in and out of the nucleus

The lamina provides mechanical support and is involved in crucial cell functions, including DNA replication, cell division, and chromatin organization.

The *nucleoplasm* is the nucleus's semifluid medium, analogous to the cell's cytoplasm. In the nucleoplasm, DNA and proteins interact to form *chromatin*.

Nucleolus location and function

The *nucleolus* is a nonmembrane-bound region in the nucleus where ribosomal RNA (rRNA) is transcribed subunits assembled.

Chromosomal loci of the ribosomal RNA (rRNA) genes are known as nucleolar organizing regions (NORs). rRNA is essential for ribosome formation during protein synthesis (i.e., translation).

The rRNA subunits are exported from the nucleolus to the cytoplasm for assembly into ribosomes to translate mRNA into proteins. The nucleolus is the site of transcription and processing of rRNA subunits.

Thus, it has DNA, RNA, and ribosomal proteins, including RNA polymerases, imported from the cytosol. Under the light microscope, the nucleolus is prominent in cells with high protein production.

Nuclear envelope and nuclear pores

Nuclear envelope (or *nuclear membrane*) is a double membrane system composed of an outer and inner membrane.

The nuclear envelope is analogous to the plasma membrane surrounding the cell.

Perinuclear space is between the two layers of the nuclear membrane.

Nuclear pores punctuate the nuclear double membrane for selective passage of specific biomolecules entering the nucleus.

Cell processes and communications require the segregation of biomolecules.

The number of nuclear pores is not static but changes based on the cell's needs.

Through the pores, signal molecules, nucleoplasm proteins, nuclear membrane proteins, lipids, and transcription factors can enter the nucleus, while mRNA, rRNA, and ribosomal proteins exit into the cytoplasm.

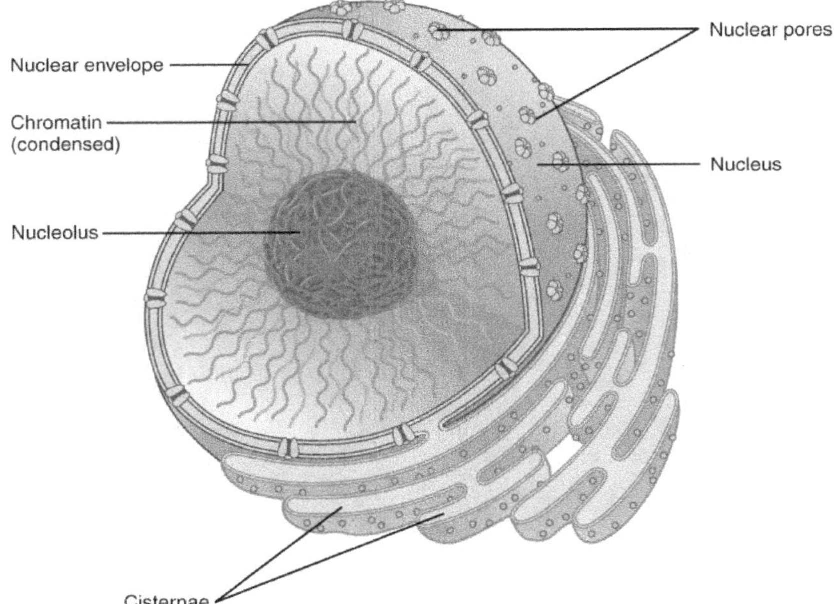

The nucleolus is within the nucleus and assembles ribosomal subunits in eukaryotic cells

Notes for active learning

Membrane-Bound Organelles

Cytoplasm and endomembrane system

The cytoplasm is the cellular material outside the nucleus and within the cell's plasma membrane. It includes the *cytosol*, the cell's fluid medium, and the *organelles*, small, usually membrane-bound subunits with specialized functions (ribosomes are not membrane-bound). Among other functions, organelles structurally support the cell, facilitate cell movement, store and transfer energy, and exchange products in transport vesicles. Mitochondria in animal cells and chloroplasts in plant cells are organelles with genetic material and replicate independently of the nucleus.

The *endomembrane system* is a series of intracellular membranes that compartmentalize the cell. Vesicles bud from the endomembrane system as transport molecules within the cell. Products synthesized in the cell pass through at least some portion of the endomembrane system.

A typical pathway through the endomembrane system is:

1. Proteins produced in rough ER (endoplasmic reticulum) and lipids from smooth ER are carried in vesicles to the Golgi apparatus.

2. The Golgi apparatus modifies these products and sorts and packages them into vesicles transported to various cell destinations (e.g., organelles or exported from the cell).

3. Secretory vesicles transport products to organelles or the plasma membrane, secreted via exocytosis.

Aside from the Golgi apparatus, smooth and rough ER, and secretory vesicles, the endomembrane system includes the membranes of lysosomes, peroxisomes, and other organelles within the cell.

While most cells have the same organelles, their distribution may differ depending on the cell's function. For example, cells that require much energy for locomotion (e.g., sperm cells) have many mitochondria; cells involved in secretion (e.g., pancreatic islet cells) have many Golgi apparatuses; and cells that primarily serve a transport function (e.g., red blood cells) lack organelles.

Structure of mitochondria

Mitochondria (singular, *mitochondrion*) are responsible for aerobic respiration, converting chemical energy into ATP (adenosine triphosphate) using oxygen. ATP is used as the primary energy source within cells. Mitochondria vary in shape; they may be long and thin or short and broad. Mitochondria can be fixed in one location or form long, moving chains. They have a double membrane, with the outer membrane separating the mitochondria from the cytoplasm.

The inner membrane has folds as *cristae*, which project into the inner fluid, the *matrix* (analogous to the cytoplasm of the cell). Between the outer and inner membrane is the intermembrane space. This region is high in protons, creating a proton gradient, which drives ATP synthesis.

The cristae are dotted with *ATP synthase* protein complexes, powered by the proton gradient (used in the electron transport chain) to transform ADP into ATP. This process is essential to producing the energy that organisms require for metabolic functions. Thus, cells with higher energy needs require more mitochondria.

Endosymbiotic theory of evolution

Mitochondria are unique because they have their genome, distinct from the genome within the nucleus. They have circular DNA, inherited exclusively from the mother, containing genes for synthesizing some mitochondrial proteins.

Mitochondria can replicate their DNA independently from the nucleus. They have ribosomes independent from the host cell's ribosomes in sequence and structure.

The unique characteristics of mitochondria support the *endosymbiosis theory* for the origin of eukaryotic cells. The endosymbiosis theory states that mitochondria were once free-living aerobic prokaryotes consumed by another cell about 1.5 billion years ago.

The prokaryote (likely a proteobacterium) became an endosymbiont within the cell, providing the anaerobic host cell with ATP via aerobic respiration. In return, the host cell provided the endosymbiont with a stable environment and nutrients.

Over time, the endosymbiont transferred most of its genes to the host nucleus, to the point that it became obligate (i.e., could no longer survive outside the host cell) and evolved into a mitochondrion.

Biologists largely accept the endosymbiosis theory. One of the most compelling evidence is that mitochondrial DNA does not encode its proteins.

Many of its genes are in the nuclear DNA; therefore, proteins must be imported into the mitochondria.

Mitochondrial DNA, ribosomes, and enzymes are similar to bacterial forms, and mitochondria replicate by a process similar to binary fission.

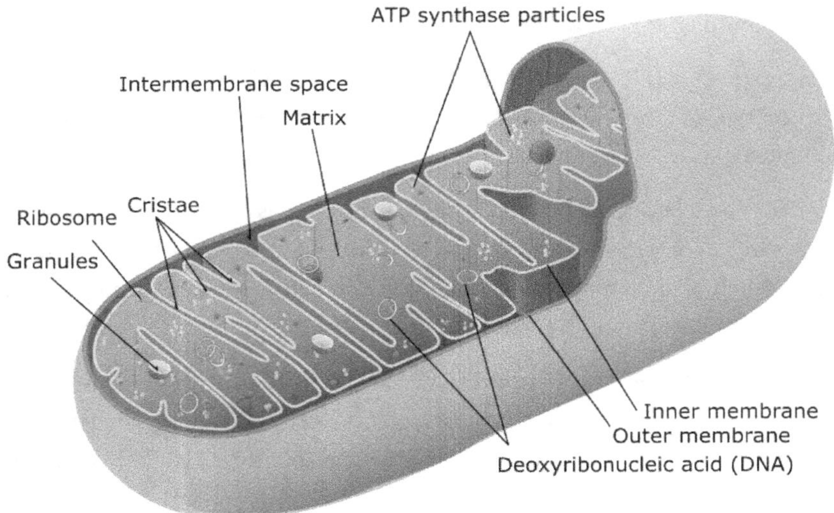

Mitochondria have double membrane enclosure with ATPase embedded in the inner membrane

Additionally, some of the proteins within the mitochondria's plasma membrane are like prokaryotes, different from proteins in the eukaryotic plasma membrane.

Chloroplasts, the organelles that conduct photosynthesis, exhibit strong evidence of an endosymbiotic origin, although they are hypothesized to have descended from cyanobacteria rather than proteobacteria. Chloroplasts and other plastids can undergo secondary and even tertiary endosymbiosis, causing the development of other membranes.

Lysosomes as vesicles containing hydrolytic enzymes

Lysosomes, only in animal cells, are membrane-bound vesicles produced by the Golgi apparatus. These small organelles contain hydrolytic enzymes (low pH) to digest macromolecules: proteins, nucleic acids, carbohydrates, and lipids. These macromolecules may originate from food, from the waste products of cells, or foreign agents, such as viruses and bacteria. After these particles enter a cell in vesicles, lysosomes fuse with vesicles and digest their contents by hydrolyzing the macromolecules into their monomers.

Lysosomes are especially important in specialized immune cells. For example, white blood cells that engulf foreign agents use lysosomes to digest the invaders.

Autodigestion is when lysosomes digest parts of the body's cells due to disease or trauma or for immune purposes (e.g., programmed cell death).

Mutations in the genes that encode lysosomal enzymes cause *lysosomal storage disorders*. When a mutation renders certain lysosomal enzymes inefficient (or inoperable), waste products accumulate in the cells and cause severe, often incurable complications.

Rough and smooth endoplasmic reticulum

Endoplasmic reticulum (ER) is a system of membrane channels (or *cisternae*) continuous with the outer membrane of the nuclear envelope. The space enclosed within the cisternae, the *lumen*, is thus continuous with the perinuclear space.

The rough ER, so-called because of its rough appearance, is studded with ribosomes on the cytoplasmic side. Here, proteins are synthesized and enter the ER interior for processing and modification. Modifications may include folding the protein or combining multiple polypeptide chains to form proteins with several subunits.

The smooth ER is usually interconnected with the rough ER but lacks ribosomes, hence its smooth appearance. It is the site of various synthesis, detoxification, and storage processes, such as synthesizing lipids and steroids and the metabolism of carbohydrates and other molecules.

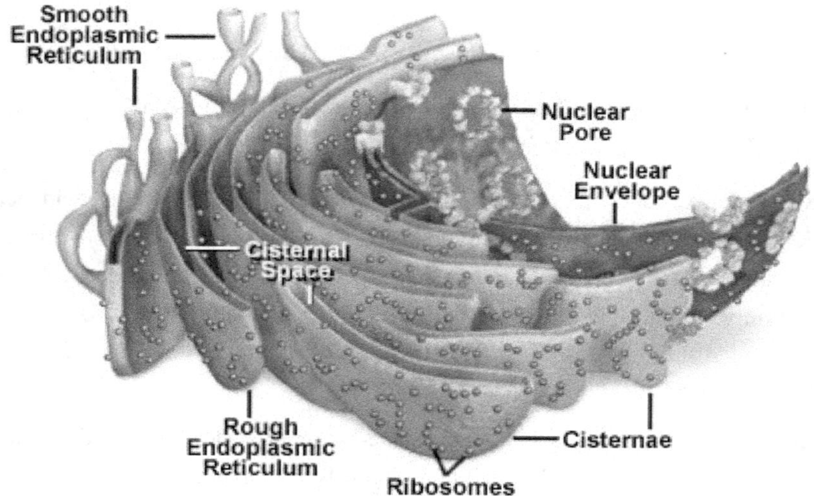

The ER forms transport vesicles for trafficking particles to the Golgi apparatus

Ribosomes

Ribosomes are organelles composed of proteins and ribosomal RNA (rRNA). They are either floating free in the cytoplasm, attached to the rough ER's surface, or within mitochondria and chloroplasts.

Ribosomes translate messenger RNA (mRNA) to coordinate the assembly of amino acids into polypeptide chains, which fold into functional proteins.

Ribosomes of eukaryotic cells are 20 to 30 nm in diameter, while those in prokaryotic cells are slightly smaller. They are composed of one large and one small subunit; each with its mix of proteins and rRNA. *Polyribosomes* are several ribosomes simultaneously synthesizing the same protein; they may be attached to ER or floating freely in the cytosol.

Smooth endoplasmic reticulum for lipids biosynthesis

The smooth ER and rough ER synthesize key membrane components. The smooth ER synthesizes the major lipids of a membrane: phospholipids, glycolipids, and steroids.

Some lipid products are already in the correct form for incorporation into a membrane once secreted by the smooth ER. Other lipids require modification by the Golgi apparatus.

Lipids synthesized in the smooth ER must pass through the Golgi apparatus before heading to their destination at the plasma membrane or membrane-bound organelles.

Rough endoplasmic reticulum for transmembrane protein biosynthesis

The rough ER synthesizes the protein components of cell membranes. This includes the plasma membrane, the membranes of the ER, Golgi apparatus, lysosomes, and other organelles.

Membrane proteins are divided into several classes (discussed later), but some of their functions include membrane transport, cell-to-cell adhesion, cell signaling, and catalysis.

Like lipids synthesized on the smooth ER, proteins synthesized on the rough ER follow a set pathway through the Golgi apparatus towards their destinations.

Transmembrane and secreted proteins with signal sequence

Proteins destined for the plasma membrane, Golgi apparatus membrane, ER membrane, or lysosomal membranes are inserted into the ER membrane immediately after synthesis on the cytosolic side of the rough ER membrane. These proteins are transported as membrane components rather than soluble proteins.

ER membrane proteins end their journey here, but the others proceed to the Golgi apparatus.

Upon post-translational processing, Golgi membrane proteins remain in the Golgi apparatus. The remaining proteins (secretory pathway) travel to the lysosome, the plasma membrane or undergo exocytosis to leave the cell.

Secretory proteins and proteins destined for the *lumen* of the ER or Golgi apparatus are released into the ER lumen following ER synthesis. ER lumen proteins remain in the ER lumen.

Golgi lumen proteins travel to the Golgi lumen, and secretory proteins travel to the Golgi and then the plasma membrane or are secreted by the cell.

Not all protein synthesis takes place on the rough ER. Free-floating ribosomes synthesize proteins designated for use in the cytosol and some organelles (e.g., nucleus, mitochondria, chloroplasts, peroxisomes) in the cytosol. After synthesis, cytosolic proteins are released directly into the cytosol.

Nuclear, mitochondrial, chloroplastic, and peroxisomal proteins are escorted to their destinations by receptor molecules.

Protein synthesis begins on free ribosomes. Therefore, proteins that need to be synthesized in the ER must be translocated there.

Posttranslational translocation to the ER occurs after a free-floating ribosome synthesizes a polypeptide.

Cotranslational translocation (common in mammalian cells) occurs as the polypeptide is synthesized.

Cotranslational translocation is facilitated by a *signal sequence* on the growing polypeptide chain. This sequence is a short chain of amino acids, mostly hydrophobic. As soon as the signal sequence emerges on the polypeptide from the ribosome, a protein-RNA complex called a *signal recognition particle* (SRP) recognizes and binds to the signal sequence and ribosome, halting translation. The SRP then targets the ribosome and polypeptide chain to the rough ER membrane, where the SRP binds to an SRP receptor.

Binding to the SRP receptor releases the SRP from the ribosome and polypeptide, allowing the ribosome itself to bind to a protein translocation complex next to the SRP receptor called *Sec61*. After binding, the signal sequence is inserted into the Sec61 membrane channel (part of the translocation complex), and polypeptide synthesis resumes.

As it grows, the polypeptide chain is translocated through the membrane channel. A *signal peptidase* enzyme cleaves the signal sequence from the rest of the polypeptide, allowing the finished polypeptide to be released into the lumen, where it undergoes folding and modification.

If the polypeptide is a membrane protein that must enter the ER membrane and not the lumen, it is inserted into the ER membrane in a variety of ways. For example, transmembrane proteins may contain a *stop-transfer sequence*, which anchors the polypeptide in the ER membrane partway through synthesis so that the polypeptide is anchored to the ER membrane rather than located in the ER lumen.

Organelles with double-membrane structures

While most organelles of the eukaryotic cell are composed of a single bilayer membrane, three key organelles have a double membrane: mitochondria, chloroplasts, and the nucleus.

Mitochondria have a double membrane structure due to their proposed evolution from an endosymbiotic prokaryote. The intermembrane space is high in proton concentration, while the matrix (like cytosol) within the mitochondria is relatively low in proton concentration. This double membrane is crucial for creating the proton gradient that drives ATP synthesis. This proton gradient powers ATP synthases, with cytochrome proteins dotting the inner membrane's cristae, combining ADP with Pi and O_2 forming ATP by oxidative phosphorylation.

Chloroplasts, the sites of photosynthesis and ATP synthesis in plant cells, have a double membrane of endosymbiotic origin, the *chloroplast envelope*. The two membranes regulate the passage of particles into and out of the chloroplast. The outer membrane is permeable to ions and metabolites, while the inner membrane is specific to transport proteins.

Unlike mitochondria, chloroplasts contain *thylakoids* as additional membrane-bound structures. The thylakoid membranes are analogous to mitochondrial inner membranes, and the spaces within the thylakoids (lumen) are analogous to the mitochondrial intermembrane space. The high proton concentration in the lumen creates the proton gradient that drives ATP synthesis on the thylakoid membranes.

A sophisticated, double-membrane nuclear envelope surrounds the nucleus. The highly selective nuclear protein pores dotting the envelope regulate gene expression by controlling the passage of transcription factors, biomolecules, and mRNA into and out of the nucleus. Since the outer membrane is continuous with the rough ER, no vesicle transport is required to transport ER proteins into the nucleus. As a result, the cell's energy requirements are lower than if transport were required.

Golgi apparatus modifies, packages and secretes glycoproteins

The *Golgi apparatus*, named for the Italian 1906 Noble laureate Camillo Golgi, consists of a stack of flattened sacs. It acts as an intermediary in the secretion of biomolecules. The Golgi apparatus receives transport vesicles from the ER and may modify their contents before packaging the protein or lipid in vesicles for transport to their destination.

Glycosylation affects a protein's structure and function and protects it from degradation. Glycosylation is when the Golgi apparatus modifies a protein by adding carbohydrates. The Golgi apparatus glycosylates proteins (i.e., add sugar residues to a protein) and modifies existing glycosylations. The finished glycosylation product is a *glycoprotein* protein with attached sugars or carbohydrates.

Peroxisomes

Peroxisomes are membrane-bound vesicles containing enzymes for a variety of metabolic reactions. They are involved in the catabolism (i.e., degradation) and anabolism (i.e., synthesis) of macromolecules, including fatty acids, proteins, and carbohydrates. When peroxisomes were first discovered, they were defined as organelles that produce hydrogen peroxide through oxidation reactions. Peroxisomes use catalase (enzymes end with ~*ase*) to degrade the produced hydrogen peroxide into water and oxygen or oxidize compounds.

Peroxisomes are abundant in the liver, where they are notable for producing bile salts from cholesterol and metabolizing alcohol. They participate in lipid biosynthesis. Peroxisomes' functions are vast and varied, making them a vital part of eukaryotic cells.

Notes for active learning

Plasma Membrane

Semi-permeable plasma membrane

The *semi-permeable plasma membrane* separates cell contents from the extracellular environment and regulates the passage of materials into and out of the cell. The plasma membrane surrounds the cell, providing support, protection, and a boundary from the outside environment. It is primarily composed of lipids and proteins, forming a dynamic bilayer of lipids with membrane proteins. The function and composition of the two layers of a plasma membrane differ; therefore, the membrane is asymmetric.

Phospholipid bilayer with embedded proteins and steroids

Lipids are a large group of naturally occurring hydrophobic molecules. They are vital components of the plasma membrane. Lipids in the plasma membrane include phospholipids, steroids, and glycolipids. The foundation of the plasma membrane is the *phospholipid bilayer*. Phospholipids are molecules with a phosphate head and two long hydrocarbon tails. The head is hydrophilic and attracts water and other polar molecules, while the hydrocarbon tails are hydrophobic and repel water molecules.

The special dual nature of phospholipids, called *amphipathic*, allows them to align into a bilayer when placed in water spontaneously. In the bilayer, the hydrophilic phosphate heads point out towards the aqueous solution, while the hydrophobic tails point inward towards one another. Thus, the plasma membrane's extracellular and intracellular surfaces are hydrophilic, while the membrane's interior is hydrophobic. Hydrophobic interactions in the interior of the membrane hold the entire structure together.

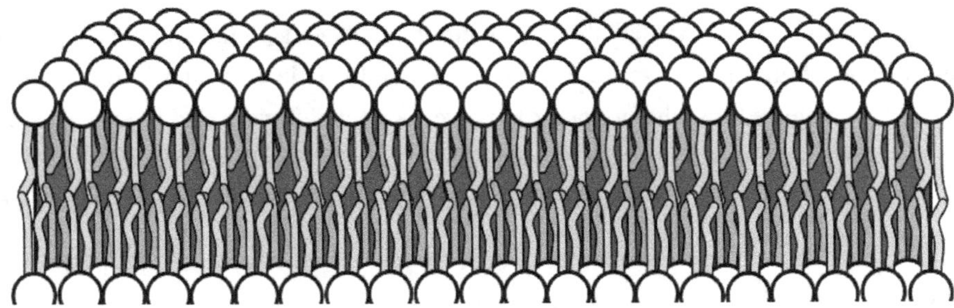

Phospholipid bilayer (absent in the schematic are embedded proteins)

The bilayer provides the plasma membrane with stability and with extraordinary flexibility. Lipids exhibit free lateral diffusion about the bilayer, resulting in varying lipid compositions across the membrane sections. Generally, cell membranes have a consistency like that of olive oil at room temperature.

Increasing the concentration of lipids with unsaturated hydrocarbon tails increases membrane fluidity; the addition of saturated hydrocarbon tails makes the membrane rigid.

Cells regulate membrane fluidity by lengthening phospholipid tails, altering the cytoskeleton, changing their protein composition, and adding steroids (e.g., cholesterol).

Steroids are a class of lipids that regulate membrane fluidity by hindering phospholipid movement. Cholesterol, a steroid in animal plasma membranes, plays a crucial role in maintaining membrane fluidity despite temperature fluctuation. At high temperatures, membranes become dangerously fluid and permeable unless cholesterol interferes with extreme phospholipid movement.

At low temperatures, membranes may freeze unless cholesterol prevents phospholipids from becoming stationary due to strong hydrophobic interactions.

Cholesterol molecules facilitate cell signaling and vesicle formation.

Glycolipids are lipids modified with a carbohydrate. In the plasma membrane, they assist in various functions and anchor the plasma membrane to the *glycocalyx*, a layer of polysaccharides linked to the membrane lipids and proteins.

Essentially, the glycocalyx is a carbohydrate coat present on the extracellular surface of the plasma membrane and the extracellular surface of the cell walls of some bacteria. The glycocalyx's carbohydrate chains face outwards, providing markers for cell recognition and adhesive capabilities to the cell.

Protein components and the fluid mosaic model

In the early 1900s, researchers noted that lipid-soluble molecules entered cells more readily than water-soluble molecules, suggesting that lipids are a component of the plasma membrane. Chemical analysis later revealed that the membrane indeed contained phospholipids. The amount of phospholipid extracted from a red blood cell was just enough to form one bilayer. The analysis suggested that the nonpolar tails were directed inward and polar heads outward. To account for the permeability of the membrane to non-lipid substances, researchers initially proposed a *sandwich model*, describing a phospholipid bilayer in between two layers of protein.

After investigation with an electron microscope, the *unit membrane model* was proposed, which was based on the "trilaminar" appearance of two dark outer lines and a light inner region visible under the electron microscope. The dark outer lines were believed to be protein monolayers, while the inner region was thought to be a phospholipid bilayer. The unit membrane model was essentially in agreement with the sandwich model. However, both of these models failed to explain permeability satisfactorily.

Fluid mosaic model and associated proteins of membranes

In 1972, Garth L. Nicholson and Seymour J. Singer published the currently accepted *fluid-mosaic model*, which describes a plasma membrane as a phospholipid bilayer embedded with proteins. Electron micrographs of the freeze-fractured membrane (and other evidence) supported the fluid-mosaic model. The lipid portion of the plasma membrane gives it its "fluid" characteristic. Thus, fluidity describes the lipids that diffuse freely throughout the membrane and regulate consistency.

The membrane's protein components contribute to the "mosaic." Protein composition in the plasma membrane depends on the function of the cell. Some proteins are held in place by cytoskeletal filaments, but most migrate within the fluid bilayer.

The proteins embedded in a membrane are grouped into two classes by location.

Peripheral membrane proteins are on the membrane surface, mainly the intracellular side, interacting with cytoskeletal elements to influence cell shape and motility.

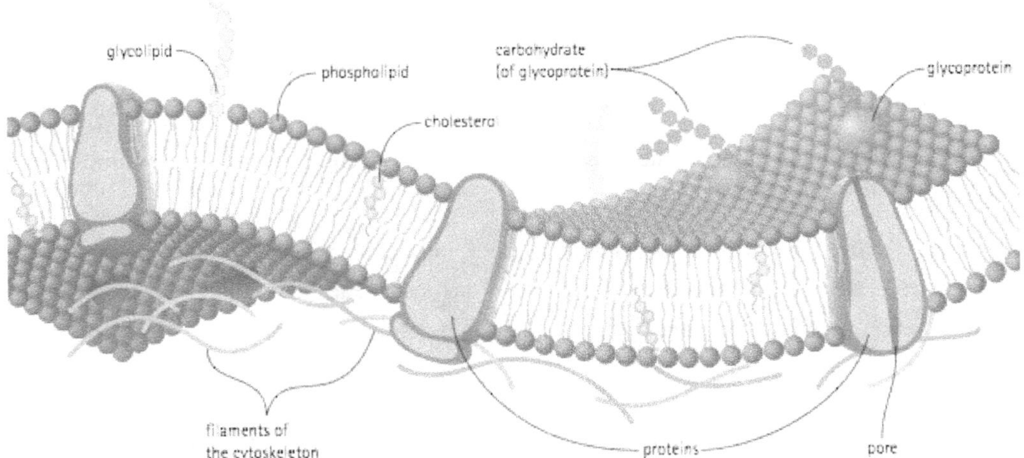

The phospholipid bilayer with cholesterol between the hydrophobic lipid tails and embedded proteins

These proteins are not amphipathic; they interact with the membrane's hydrophilic heads.

Peripheral membrane proteins are removed from the membrane with relative ease using high salt or high pH, and therefore are not permanently attached to the membrane.

Integral membrane proteins are permanently attached to the membrane and cannot be removed without disrupting the lipid bilayer. They possess hydrophobic domains anchored to hydrophobic lipids.

Most integral membrane proteins are *transmembrane* proteins, spanning the entire membrane.

Membrane proteins participate in cell signaling, cell-to-cell adhesion, transport through the membrane, enzymatic activity, and other biochemical activities.

When divided by function, several key classes of membrane proteins emerge.

Functional classes of membrane proteins

Receptor proteins provide a binding site for hormones, neurotransmitters, and other signaling molecules. Receptor proteins are usually specific in that they bind to a single molecule or class of molecule. The binding of the signal molecule to the receptor triggers a cellular response corresponding to a specific biochemical pathway.

Adhesion proteins attach cells to neighboring cells for cell-to-cell communication and tissue structure. These proteins generally attach to one cell's cytoskeleton and extend through the plasma membrane to the extracellular environment, where they bind and interact with the adhesion proteins of another cell.

Transport proteins move materials into and out of the cell.

These include *channel proteins* and *carrier proteins.*

Channel proteins provide a passageway for large, polar, or charged molecules that cannot pass through the lipid bilayer without assistance.

The channel proteins facilitate the passive transport of molecules; they do not require ATP to operate.

However, carrier proteins may facilitate passive and active (energy-requiring) transport of molecules. They bind to specific molecules on one side of the cell membrane and changes conformation to release the molecule on the other side of the membrane.

Enzymatic membrane proteins carry out metabolic reactions at the cell membrane.

For example, enzymatic membrane proteins help digest membrane components for recycling. The mitochondrial membrane contains enzymatic proteins (e.g., protein complexes of the electron transport chain in aerobic cellular respiration).

Recognition proteins are glycoproteins that identify a cell to the body's immune system. They allow immune cells to recognize a substance as belonging to the body or as an invasive foreign agent to be destroyed.

Antigens are the basis for A, B, and O blood groups in humans.

Immune cells recognize the sugars attached to these proteins and attack any red blood cells with foreign sugars, so patients of blood groups cannot donate blood or receive blood from people with other blood groups.

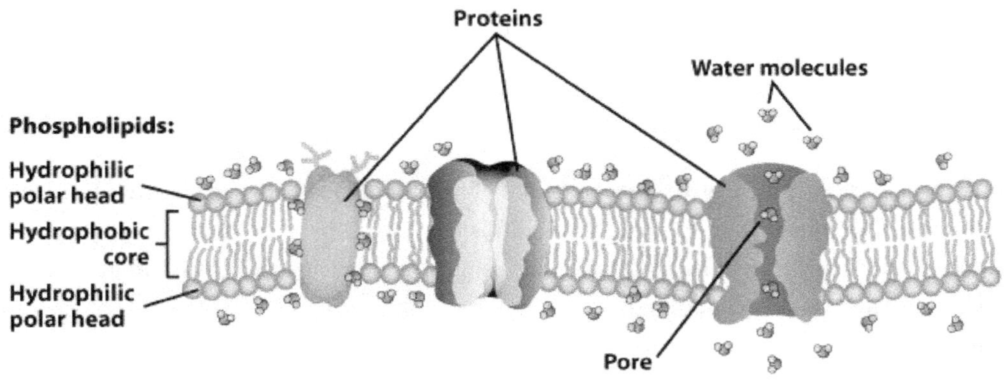

Semi-permeable membranes produce osmotic pressure

All fluids of the body are *solutions*: they contain dissolved substances (solutes) and a fluid (water) in which the substances dissolve (solvent).

Diffusion is the movement of solutes from an area with a higher to an area with a lower concentration.

Osmosis is the diffusion of water from a low solute concentration to an area of high solute concentration. Solutes diffuse to an area of lower solute concentration, while water (a solvent) diffuses to an area of higher solute concentration. The solute and solvent offset unequal solute concentration and restore equilibrium.

The natural inclination of solutes is to diffuse until they are evenly distributed. However, areas of the cell/tissue/organ/body require different solute concentrations. Separation areas are accomplished via the complex system of membrane compartmentalization in organs, tissues, and cells.

The body must prevent body compartments from reaching equilibrium but must allow the passage of certain atoms, ions, and molecules through membranes to the areas needed. Membranes are highly selective and tightly regulated.

Colligative properties of solutions depend on the concentration of a solute in the solution. They do not depend on the chemical nature of the solutes. The most important colligative property for the regulation of body fluids is *osmotic pressure*, the pressure that must be applied to prevent the net flow of water into a solution separated by a membrane (i.e., a measure of water's tendency to flow from one solution to another). This property is related to several other key terms in osmoregulation (the regulation of osmotic pressure).

Osmolarity is the total solute concentration of a solution measured in *osmoles*. Osmolarity considers penetrating and non-penetrating solutes and describes a single solution or compares different solutions.

A *hyperosmotic* solution has a higher osmolarity.

A *hypoosmotic* solution has a relatively low osmolarity.

Isosmotic solutions have the same osmotic pressure.

Tonicity describes the relative concentration of two solutions separated by a selectively permeable membrane and explains how diffusion occurs between them (e.g., between intracellular and extracellular fluid). Unlike osmolarity, tonicity refers to non-penetrating solutes (solutes that cannot cross a membrane) and describes how one solution compares to another.

A *hypertonic* solution has a relatively higher concentration of solute. Conversely, a *hypotonic* solution has a relatively lower concentration of solute. Therefore, a cell with a lower concentration of solute than the extracellular fluid is hypotonic to the fluid, while the extracellular fluid is hypertonic to the cell.

The reverse is true for a cell with a higher concentration of solute than the extracellular fluid.

A cell placed in a hypertonic solution shrinks through a process of *crenation* (i.e., shrinking of the cell) as water diffuses out of the cell to offset the high concentration of the outside solution. Conversely, a cell in a hypotonic solution swells as water rushes into the cell, causing *cytolysis* (i.e., bursting of a cell due to osmotic imbalance caused by excess water entering the cell.

Too much water can enter the cell, resulting in *lysis* (breakage of the cell).

An *isotonic* solution has an equal solute concentration to the solution it is being compared to. In this situation, there is no net water movement.

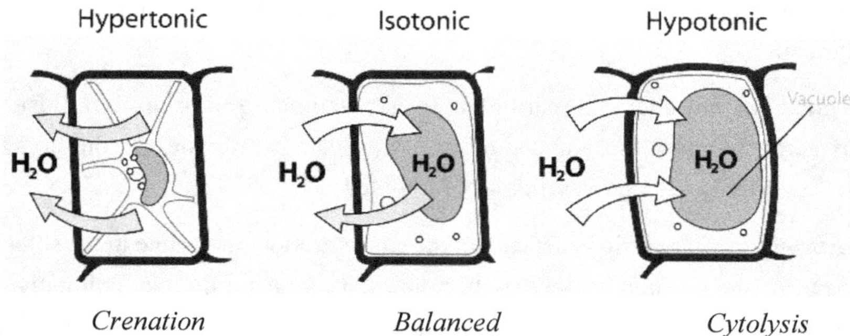

Crenation   *Balanced*   *Cytolysis*

Passive transport, diffusion and osmosis

The plasma membrane is selectively permeable; specific molecules can pass through. A molecule's ability to diffuse through the plasma membrane depends on the molecule's size, charge, and polarity. The greater the diffusing particle's lipid solubility, the easier it passes through the membrane.

Generally, smaller particles diffuse more rapidly than larger ones, and hydrophobic solutes diffuse faster than hydrophilic solutes.

Many particles cannot diffuse through the plasma membrane without assistance. Small, non-charged, or non-polar molecules pass through the membrane freely. Large, charged, or polar molecules usually require assistance to pass through the membrane.

Passive transport enables the movement of molecules across a membrane without energy expenditure by the cell. The methods include simple diffusion, osmosis, and facilitated diffusion. Passive transport utilizes a *concentration gradient*, whereby particles diffuse from an area of higher to an area of lower solute concentration.

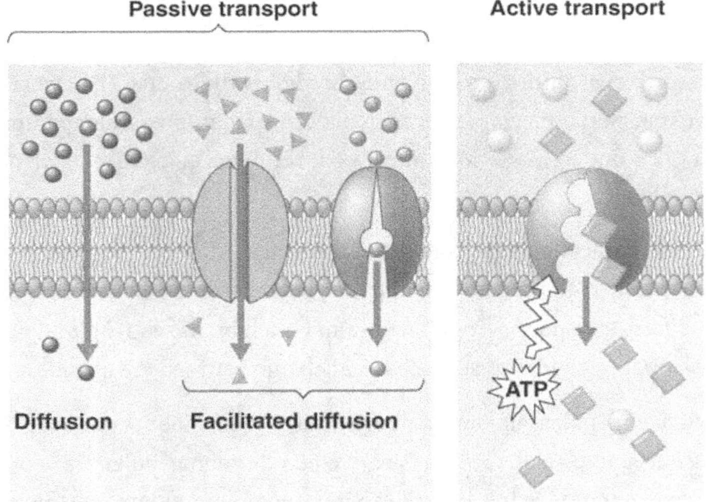

Passive transport includes diffusion and facilitated diffusion with the expenditure of energy as molecules move down the concentration gradient

Simple diffusion is the process for smaller, lipid-soluble molecules to diffuse through the phospholipid bilayer. For example, oxygen and carbon dioxide pass through the membrane via simple diffusion.

While water is a polar molecule, it is small enough to diffuse freely across a plasma membrane. *Osmosis* is the passive diffusion of water molecules.

Osmosis occurs when water moves from a region of lower solute concentration to a higher solute concentration region and is facilitated by *aquaporins* as channel proteins. Osmosis is classified as simple diffusion, despite requiring transport proteins like facilitated diffusion.

Facilitated diffusion allows larger, lipid-insoluble molecules that cannot freely pass through the phospholipid bilayer (e.g., sugars, ions, and amino acids) to get transported down their concentration gradient with the assistance of a carrier protein or channel proteins, often as a *uniporter.*

ATP needed for primary active transport

Active transport requires cellular energy to move solutes against their concentration gradient. Unlike passive transport, which exploits molecules' natural inclination to move down their concentration gradient, active transport requires energy expenditure to resist this opposing force. Carrier proteins are the transmembrane proteins that mediate molecules' movement too polar or too large to move across a membrane by diffusion, thereby governing active transport.

During this process, a solute (molecule to be transported) binds to a specific site on a transporter on one surface of the membrane. The transporter changes shape to expose the bound solute to the opposite side of the membrane. The solute dissociates from the transporter and is on the opposite side from which it started.

There are two types of active transport: primary and secondary.

Primary active transport utilizes energy generated directly from ATP. The carrier proteins for primary active transport are pumps. An example is a sodium-potassium pump, which works by moving 3 Na^+ ions out and 2 K^+ ions into a cell, resulting in a net transfer of positive charge outside the membrane.

For a cell at rest, intracellular K^+ concentration is high, and Na^+ concentration is low, while extracellular K^+ concentration is low and Na^+ concentration is high. These concentration gradients facilitate transport across the plasma membrane and help the cell manage its *membrane potential*, the difference in electrical charge between the outside and inside the cell.

Cellular sodium and potassium concentrations are maintained via active transport by the sodium-potassium pump (Na^+ / K^+ ATPase). When the intracellular Na^+ concentration is too high, and K^+ concentration is too low, the sodium-potassium pump must pump Na^+ out of the cell and K^+ into the cell to restore the appropriate concentration gradients.

Energy coupling drives secondary active transport

The energy for *secondary active transport* comes from an electrochemical gradient established by primary active transport. Secondary active transporters work via a mechanism of *cotransport.*

Cotransport occurs when one molecule moves with (down) its concentration gradient, while another molecule moves against (up) its concentration gradient.

The energetically favorable movement of the molecule moving with its concentration gradient powers the other molecule's movement against its concentration gradient.

Antiporters (e.g., sodium-calcium exchanger) move molecules in opposite directions, i.e., one is transported out while the other is transported into the cell.

Symporters move both molecules in the same direction.

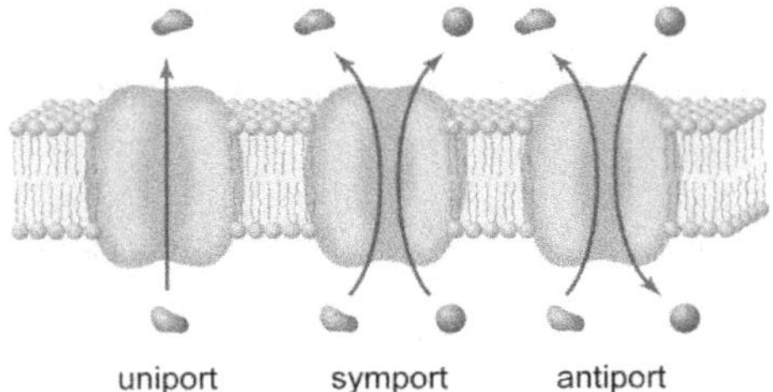

One example of a symporter is the sodium-glucose linked transporter (SGLT), which transports sodium *with* its concentration gradient from the exoplasmic space to the cytoplasmic space, and transports glucose *against* its concentration gradient from the exoplasmic space to the cytoplasmic space. These movements are energetically coupled. Note that while both molecules are moving in the same *physical* direction in a symporter, the molecules are moving in opposite directions but are energetically favorable with one process driving the other.

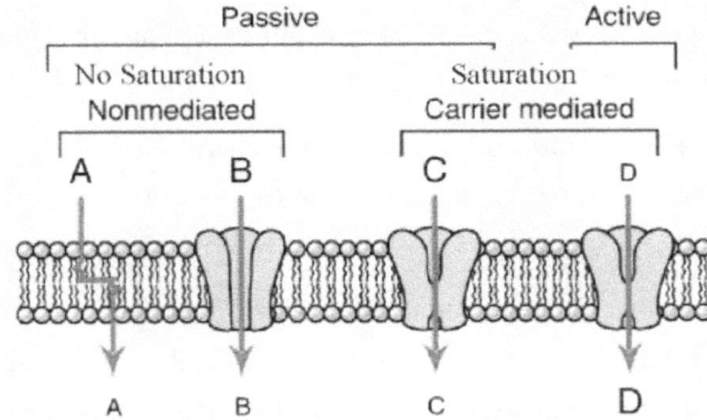

Comparison of passive and active transport: A is diffusion with movement unaided by a protein, B is passive transport down the concentration gradient through a channel (no saturation), C is passive transport through a carrier protein (saturation), D is active transport up a concentration gradient (saturation possible and energy needed)

Membrane channels

Membrane channels are transmembrane proteins that allow ions to diffuse across the membrane via passive transport.

Individual cells have different permeabilities, depending on their membrane channels.

The channel's diameter and the polar groups on the protein subunits forming channel walls determine the permeability of the channels for various ions and molecules.

Porins are channel proteins less chemically specific than many other channel proteins; generally, if a molecule can fit through the porin, it can pass through it.

Ion channels allow for the passage of ions.

Channel gating is the opening and closing of ion channels to the molecules they transport.

Changes in membrane potential modulate voltage-gated channels.

Ligand-gated channels are modulated by allosteric or covalent binding of ligands to the channel protein.

Ligands are small molecules that bind to a protein or receptor, usually to trigger a signal.

Mechanically gated channels are modulated by mechanical stimuli such as stretching, pressure, or temperature.

Several factors influence a single channel, and the same ion may pass through several channels.

Electrochemical gradient produces a membrane potential

The membrane potential is the electrical potential difference between the intracellular and extracellular environments. Nearly all eukaryotic cells maintain a non-zero membrane potential. The membrane potential is mediated by channels and pumps, altering electrochemical gradients as needed by the cell.

Whenever there is a net separation of electric charges across a cell membrane; a membrane potential exists for all cells. However, it is often the focus of neurons and depolarization with action potentials.

The concentration gradient influences molecules, but differences influence ions, creating the resting membrane potential. The *electrochemical gradient* is the combined forces of membrane potential and concentration gradient. These forces may oppose one another, work independently, or work in conjunction.

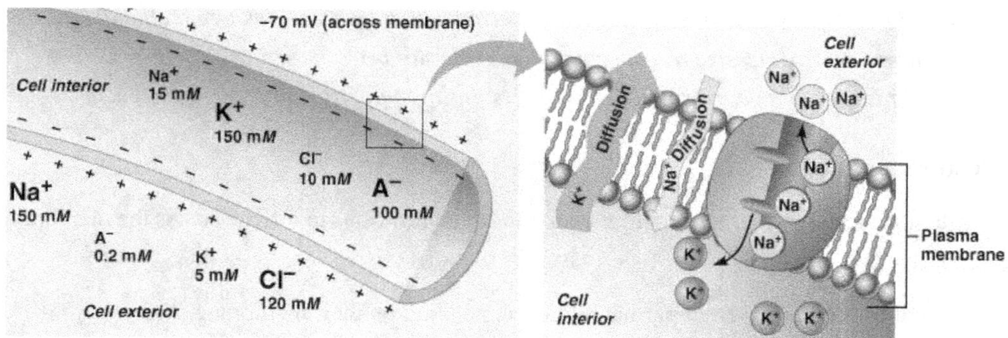

The differences in ion concentrations maintain membrane potential (i.e., voltage)

Nearly all eukaryotic cells maintain a non-zero membrane potential. In animal cells, this value ranges from −40 mV to −80 mV; thus, with respect to the extracellular environment, the inside of the cell has a negative voltage. Membrane potential is especially important for neurons, which have a *resting membrane potential* of about −70 mV. Changes in this resting potential allow for the electrical communications of neurons.

Membrane receptors and cell signaling pathway

Cells must communicate with their neighbors as well as their environment.

Cell signaling is the system by which cells receive, integrate, and send signals to communicate information.

Signals are sent through the *extracellular matrix*, a collection of polysaccharides and proteins secreted by cells in multicellular organisms.

This matrix fills the space between cells, providing structure, facilitating cell signaling, and allowing cells to move and change their shape.

The composition of the extracellular matrix is highly variable depending on tissue type, as different tissues have different functions. For example, the extracellular matrix of bone is highly calcified, while the extracellular matrix of blood is fluid, containing dissolved proteins and other molecules. The most extensive extracellular matrices are of connective tissues, which are found throughout the body and include bone and blood. Connective tissues are largely composed of their extracellular matrix and are only sparsely populated with cells.

A typical connective tissue extracellular matrix includes fibrous proteins and *proteoglycans*, glycoproteins that form a packing gel around the fibrous proteins. The primary fibrous proteins are *collagen* and *elastin*, which provide structure and flexibility, along with *fibronectin* and *laminin*, which assist in adhesion and cell migration. Transmembrane proteins such as *integrins* bind to specific proteins in the extracellular matrix and membrane proteins on adjacent cells. Integrins help organize cells into tissues. They are responsible for transmitting signals from the extracellular matrix to the cell interior.

Membrane receptors may be on the plasma membrane or intracellular membranes.

Not all signal molecules bind to membrane receptors at the plasma membrane. For example, steroid hormones and gases diffuse across the plasma membrane and bind to intracellular receptors.

Other signal molecules enter the cell via endocytosis. In addition to hormones and gases, signal molecules include neurotransmitters, proteins, and lipids.

Membrane receptors are crucial in the cell-signaling pathways. When a signal molecule binds to a membrane receptor, it may initiate a metabolic response, change the membrane potential or alter gene expression.

Signal transduction is when one signaling molecule triggers a multi-step chain reaction, which indirectly transmits the initial signal to its destination. *Second messengers* are the molecules that relay the signal.

At steps in the signal transduction pathway, second messengers can greatly amplify the strength of the original signal by increasing the number of molecules they activate. For example, one signal molecule at the membrane receptor may produce 10 second messengers, and these 10 second messengers may each produce another 10 second messengers, and so on, amplifying the signal significantly.

Cell signaling is a complex process but can be divided into four general categories. *Endocrine signaling* is when a cell secretes a signal that travels to a distant target cell. *Paracrine signaling* is when the target cell is nearby, but not in direct contact. *Juxtacrine signaling* is the signaling of a target cell in direct contact with the secreting cell. *Autocrine signaling* targets the same cell that secreted the signal.

Exocytosis and endocytosis

The plasma membrane's fluidity allows it to change shape, pinch off, and reform. This fluidity enables substances to exit the cell via *exocytosis* and enter the cell via *endocytosis*.

Both processes require cellular energy and are therefore a form of active transport.

During exocytosis, membrane-bound vesicles in the cytoplasm fuse with the plasma membrane, and their contents are released outside the cell. The vesicle assimilates into the plasma membrane, replenishing portions of the membrane that would otherwise be lost.

The exocytosis process is triggered by stimuli, leading to an increase in cytosolic calcium concentration, which activates proteins required for the vesicle membrane to fuse with the plasma membrane. Exocytosis provides a route for releasing synthesized proteins as either extracellular secretions or proteins and lipids destined for the plasma membrane.

Endocytosis is essentially the opposite process of exocytosis. During endocytosis, extracellular molecules destined for the plasma membrane or the cytoplasm are imported into the cell.

In preparation for endocytosis, a region of the outer side of the plasma membrane invaginates to enclose the imported material. This indentation folds and pinches into a membrane-bound vesicle inside the cell.

Pinocytosis ("cell-drinking" or fluid endocytosis) is a form of endocytosis that cells use to import small amounts of extracellular fluid containing molecules absorbed by the cell. The process of pinocytosis may be non-specific or mediated by receptors on the plasma membrane.

A few types of specialized cells perform phagocytosis.

Phagocytosis (or *cell-eating*) involves importing larger particulate matter than imported by pinocytosis.

The particulate matter must be degraded before it is absorbed by the cell. As such, phagocytes digest bacteria, viruses, and cell debris as a function of our immune system.

Some unicellular eukaryotes, such as amoeba, rely on phagocytosis for nutrient intake.

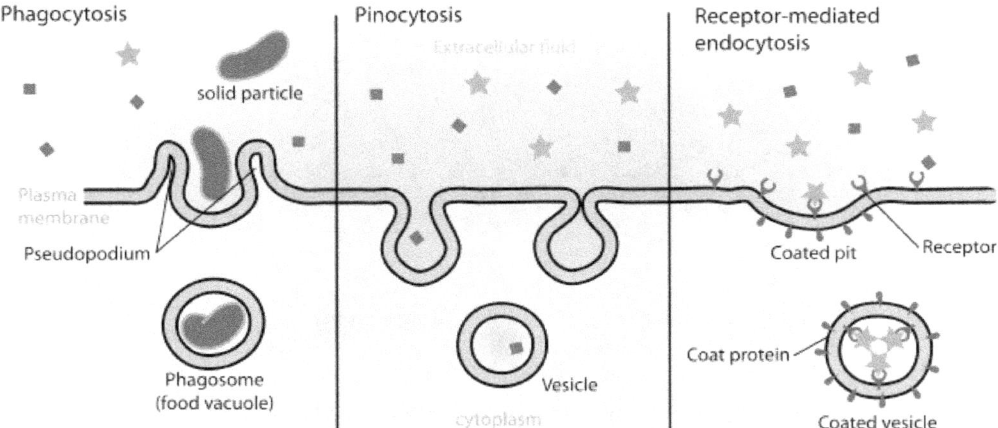

Types of endocytosis with substrates taken into the cell

Gap junctions, tight junctions and desmosomes

Cell junctions are points of contact that physically link neighboring cells.

Animal cells have three intercellular junctions: gap junctions, tight junctions, and anchoring junctions.

Gap junctions are protein channels that link the cytoplasms of adjacent cells. Gap junctions are communicating junctions because they allow for rapid cell-to-cell communication. They form by joining two membrane channels on adjacent cells, allowing small molecules and ions between cells while still preventing their cytoplasms from mixing.

Gap junctions are essential in cardiac muscle tissues, where electrical impulses must be transmitted through cells exceptionally rapidly so that the muscle fibers contract as a single unit.

The *tight junction* has plasma membrane proteins attach in zipper-like fastenings, holding cells so tightly that the tissues become barriers to molecules.

Tight junctions are formed by the physical joining of the extracellular surfaces of two adjacent plasma membranes, producing a seal that prevents the passage of materials between cells.

Materials must enter the cells by passive or active transport to pass through the tissue.

Tight junctions are essential in areas where more control over tissue processes is needed. For example, the epithelial cells in the intestine involved in nutrient absorption.

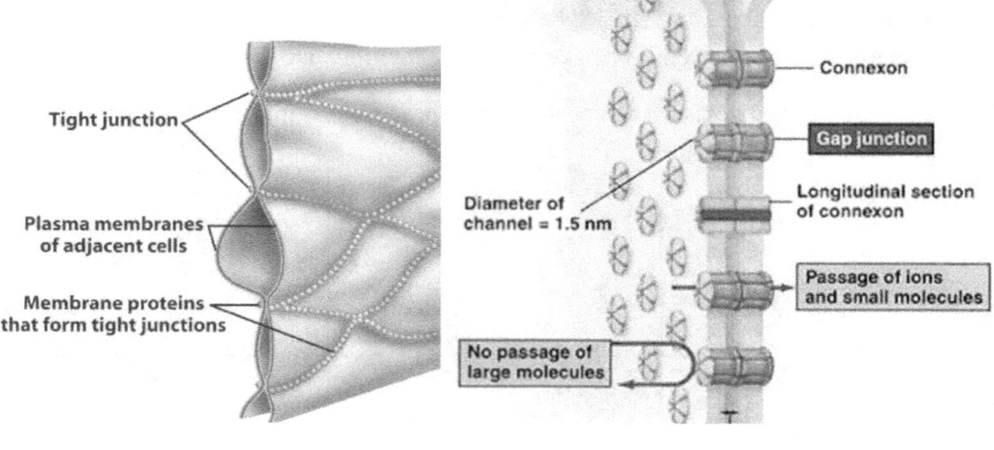

Tight junctions   *Gap junctions*

Anchoring junctions use proteins extended through one cell's plasma membrane and attached to another cell. Anchoring junctions are firm but still allow for spaces between adjacent cells.

Desmosomes, an anchoring junction, are created by dense protein patches on the plasma membranes of two cells. Internally, proteins anchor to the cell's cytoplasm, while the proteins adhere to one another.

The purpose and function of desmosomes are to hold adjacent cells firmly in place in tissue areas subject to stretching (e.g., bladder, skin, stomach).

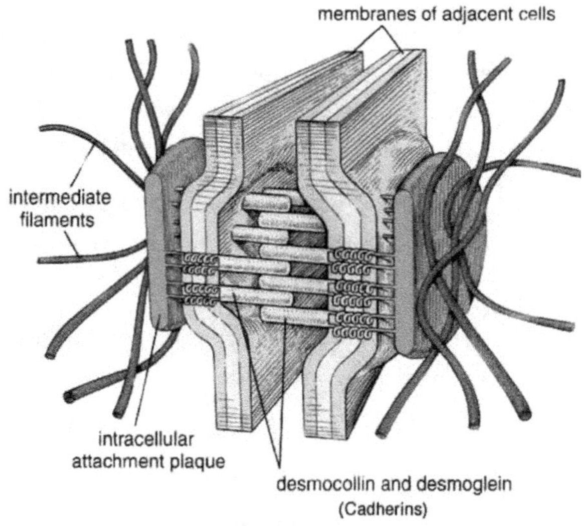

Desmosome

Cytoskeleton

Cytoskeleton for cell support and movement

The *cytoskeleton* is a scaffold of flexible, tubular protein fibers extending between the nucleus to the plasma membrane in eukaryotes. This vast network of fibers maintains the shape of the cell, provides support, and facilitates vesicular transport.

The cytoskeleton is the cellular analogy to an animal's bones and muscles. It anchors organelles and enzymes to cell regions to keep them organized in the cytosol.

The cytoskeleton changes shape to facilitate contractility and movement, allowing the cell to divide, migrate or undergo endocytosis and exocytosis.

During cell division, the cytoskeletal elements rapidly assemble and disassemble, forming spindles for chromosomes' organization and cleaving the cell into two daughter cells.

The cytoskeleton's long fibers serve for intracellular transport, upon which vesicles and organelles move via motor proteins (e.g., dynein and kinesin).

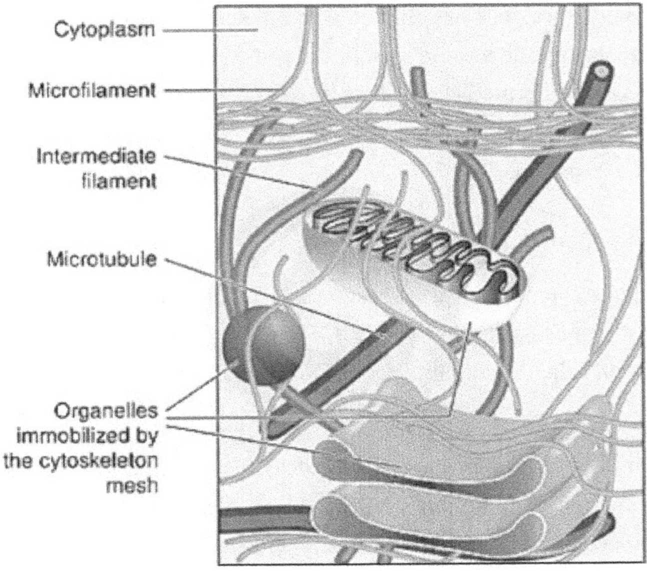

The cytoskeleton is a network of tubular proteins to provide shape and facilitate the transport of vesicles

Microfilaments for cleavage and contractility

Microfilaments (i.e., actin filaments) are the thinnest and abundant of the cytoskeleton proteins. They are contractile protein *actin* composed of long, thin fibers, about 7 nm in diameter, in bundles or mesh-like networks. Microfilament consists of two globular actin subunits chains twisted to form a helix.

Microfilaments assemble and disassemble quickly according to the needs of the cell. They are involved in cell motility functions, such as muscle cells' contraction, the formation of amoeba pseudopodia, and cleavage of the cell during cytokinesis. While flexible, microfilaments are strong and prevent deformation of the cell by their tensile strength.

Microfilaments provide tracks for the movement of myosin. Myosin attaches to vesicles (or organelles) and pulls them to their destination along the microfilament track. Additionally, the interaction between microfilaments and myosin is crucial to the cell's function.

Intermediate filaments for support

Intermediate filaments are thicker than microfilaments but thinner than microtubules. Typically, they are 8 to 11 nm in diameter. These rope-like assemblies of fibrous polypeptides are found extensively in regions of cells subjected to stress. Most intermediate filaments are in the cytoplasm, supporting the plasma membrane and forming a cell-to-cell junction. However, *lamins* are a class of intermediate filaments responsible for structural support within the nucleus. Unlike microfilaments and microtubules, intermediate filaments cannot rapidly disassemble once assembled.

Microtubules for support and transport

Microtubules (tubulin) are hollow protein cylinders about 25 nm in diameter and 0.2–25 μm in length. They are the thickest and most rigid of filaments. Microtubules are globular proteins of *tubulin* and β tubulin. The microtubule assembly combines these as dimers, and the dimers are arranged in rows.

Microtubule strength and rigidity make them ideal for resisting the compression of the cell. However, these fibers serve functions similar to microfilaments. Microtubules act as tracks for intracellular transport, but they interact primarily with *kinesin* and *dynein* motor proteins rather than myosin.

Microtubules' transport function is crucial for trafficking neurotransmitters throughout nerve cells. Like microfilaments, microtubules are rapidly assembled or disassembled. Regulation of microtubule assembly is under the control of a *microtubule-organizing center* (MTOC). Microtubules radiate from the MTOC and extend throughout the cytoplasm. During cell division, the centrosome generates the microtubule spindle fibers necessary for chromosome separation.

Eukaryotic cilia and flagella

Cilia and *flagella* are two microtubule complexes that protrude from the cell body and serve cell motility and sensory functions. Eukaryotes are membrane-bounded cylinders that enclose a matrix of nine pairs of microtubules encircling two single microtubules, a *9 + 2 pattern*. Movement occurs when these microtubules slide past one another. A basal body anchors the cilium (or flagellum) to the cell body at the plasma membrane. The basal body is derived from a centriole formed by nine pairs of microtubules without central microtubules, the *9 + 0 patterns*. Eukaryotic cilia and flagella grow by polymerizing (adding) tubulin to their tips.

Cilia are short, hair-like projections. Nearly every human cell has at least one cilium, *non-motile,* and functions as a sensory antenna important in cell signaling pathways. Non-motile cilia lack the 2 central microtubules and have a 9 + 0 pattern.

Many cells are covered with *motile* cilia, which undulates to transport particles across the cell surface. For example, motile cilia in the respiratory tract push mucus and irritants out of the lungs, trachea, and nose.

Eukaryotic flagella are structurally like eukaryotic cilia, so distinctions are often not drawn between them. Like cilia, they have sensory and motility functions. However, eukaryotic flagella tend to be longer and move with a whip-like motion. An example of a eukaryotic flagellum is the tail of a sperm cell.

Prokaryotic flagella are notably different from their prokaryotic cilia counterparts. Prokaryotic flagella are noted for their long, whip shape and functional differences from cilia.

Centrioles and the microtubule-organizing centers

The *centrosome* is the main microtubule-organizing center at the poles during mitosis and meiosis.

Centrosomes contain a pair of barrel-shaped organelles of *centrioles.*

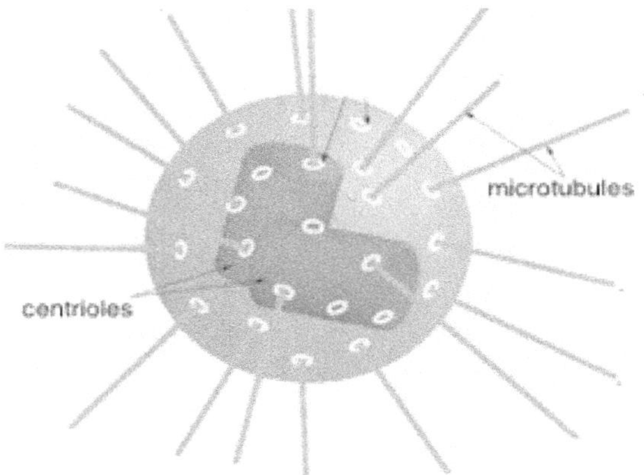

Centrosome is the microtubule organizing center with microtubules attached to the centrioles

The centrosome has a crucial role in mitosis when it divides into two centrosomes, interacting with chromosomes via microtubules to form the mitotic spindle

Centrioles are components of the centrosome. Microtubules radiate from these barrel-shaped structures, made of microtubules themselves. They are short cylinders with a ring pattern (9 + 0) of microtubule triplets.

In animal cells and most protists, a centrosome contains two centrioles oriented at right angles.

In mitosis, the terms centrioles and centrosomes are often used interchangeably because centrioles form the crucial parts of a centrosome.

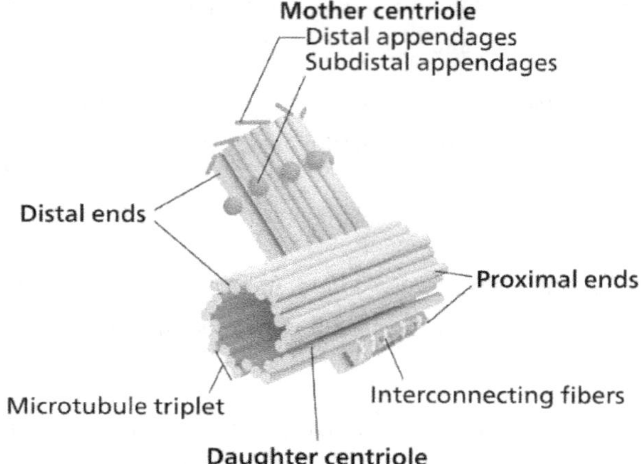

Centrosome contains two centrioles oriented at right angles

Cell Cycle and Mitosis

Methods for cell division

The cell cycle describes a cell's lifetime, detailing the life stages from creating a cell to its division into two daughter cells.

Most of an organism's cells divide throughout their lifetime (i.e., an exception is nerve cells), as cell division is the process by which organisms grow, repair tissues, and reproduce.

Mitosis and meiosis are two types of cell division.

Mitosis is cell division when new *somatic* (or *body*) cells are added to multicellular organisms as they grow and when tissues are repaired or replaced.

As a cell prepares for mitosis, it grows larger, the number of organelles doubles, and the DNA replicates.

Meiosis produces gametes (reproductive cells of egg and sperm) by organisms that reproduce sexually. Meiosis is discussed separately. .

Mitosis does not introduce genetic variations.

A daughter cell is identical in chromosome number and genetic makeup to the parent cell.

Mitosis distributes identical genetic material to two daughter cells.

What is remarkable is the fidelity with which the DNA is passed along, without dilution or error, between generations.

Eukaryotes divide by mitosis, but prokaryotes undergo *binary fission,* a form of asexual reproduction.

Binary fission is a simpler process wherein the parent cell replicates its DNA and then divides. Since prokaryotes have a single and circular DNA without a nucleus, there are no complex steps of chromosome formation and separation by mitosis with microtubules between the centromere and centrioles as in eukaryotic cells.

Interphase as the common phase of the cell cycle

Interphase is the stage before mitosis and represents most of the cell's life. This is not a static state but rather a progression towards mitosis.

However, in some cases, a cell halts its progress through interphase, either temporarily or permanently.

This may be because it is a non-dividing cell (e.g., nerve cell) or not healthy enough to perform growth and replicate interphase functions.

During interphase ($G_1 \rightarrow S \rightarrow G_2$), the cell prepares to divide by growing, replicating DNA and organelles, and synthesizing mRNA and proteins.

When it has completed interphase functions, the cell exits interphase and enters mitosis.

Four stages of mitosis

The four phases of mitosis: prophase, metaphase, anaphase, and telophase (PMAT).

1. Prophase = *Prepare*: cell *prepares* for mitosis

2. Metaphase = *Middle*: chromosomes align in the *middle* of the cell

3. Anaphase = *Apart*: centromere splits

    The sister chromatids pull *apart* by microtubules to the opposite poles

4. Telophase = *Two*: two daughter nuclei reform with separate nuclei

Prophase, the first phase of mitosis, involves chromatin condensation, nucleolus dissolution, nuclear membrane fragmentation, and centrosome movement. *Chromatin condensation* is the process by which loose euchromatin condenses into chromosomes. Currently, the chromosomes have no fixed orientation in the cell.

Upon chromatin condensation, the nucleolus dissolves, and the nuclear membrane begins to fragment, exposing the chromosomes to the cytoplasm of the cell. Simultaneously, centrosomes begin to migrate to opposite sides of the cell. Microtubules start to extend from the centrosomes, forming the *spindle apparatus*.

Metaphase follows prophase when nuclear fragmentation completes, and the centrosomes are at opposite poles of the cell. Microtubules emerge from the centrosomes and attach to the chromosomes.

Microtubules align chromosomes along an imaginary line in the center of the cell, the *metaphase plate* (or *equatorial plate*) completing the formation of the spindle apparatus. Chromosomes must be attached and aligned at the metaphase plate before the cell proceeds dividing.

During *anaphase, sister chromatids* (i.e., chromatin strands replicated during the S phase of interphase but attached at the centromere) are pulled apart to opposite poles of the cell.

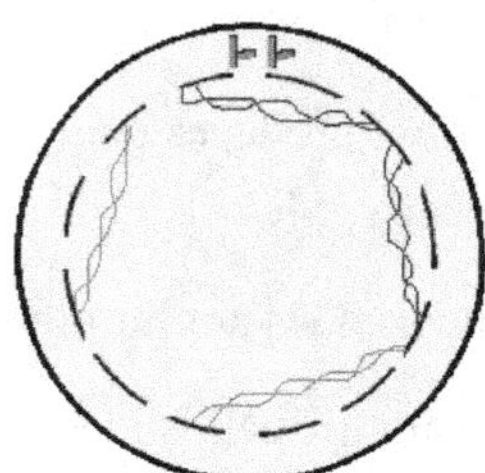

Prophase

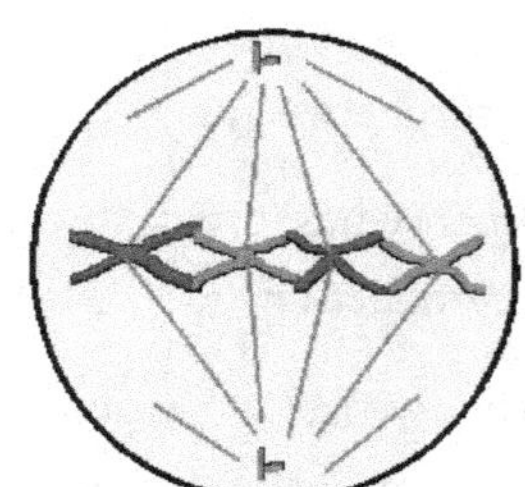

Metaphase

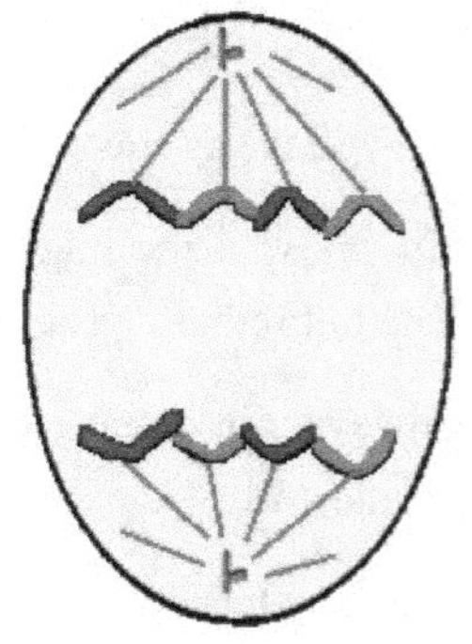

Anaphase

The sister chromatids separate at their centromere and travel towards opposite centrosomes by the spindle microtubules. By the end of anaphase, equal numbers of sister chromatids are stationed by both centrosomes.

Telophase is the cell reverse action of prophase as it prepares to divide.

The spindle apparatus disassembles, and two daughter nuclei reform.

As spindle microtubules disassemble, two regions of identical chromatids are present.

A nuclear envelope develops around each region, forming two daughter nuclei.

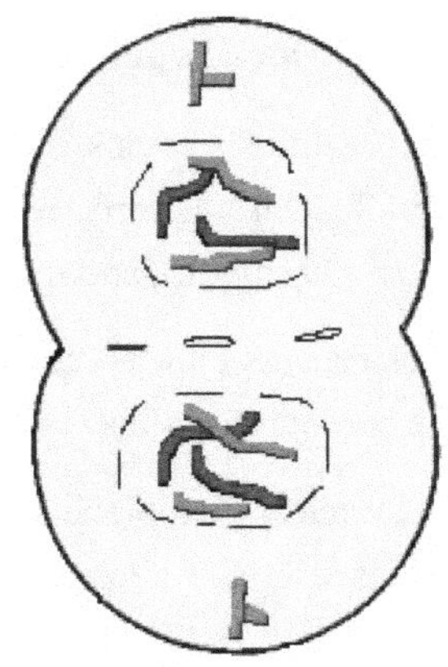

Telophase

Within the nuclei, chromosomal DNA uncoils into chromatin, and nucleoli form. The cell contains two identical daughter nuclei, and cell division proceeds.

Cytokinesis divides the cytoplasm to produce two daughter cells

Cytokinesis is the division of the cytoplasm to create two daughter cells. This usually coincides with the end of telophase but is *not* a phase of mitosis. Instead, it is a separate event that does not always occur.

When mitosis occurs but cytokinesis does not, multinucleated cell form; (often in plants, but in skeletal muscle cells of animals).

In animal cells, cytokinesis occurs by the process of *cleavage.*

First, a *cleavage furrow*, a shallow groove between the two daughter nuclei, appears. The cleavage furrow deepens as a microfilament band; the *contractile ring* constricts between the two daughter cells. A narrow bridge exists between daughter cells during telophase until constriction separates the cytoplasms.

The result is two daughter cells enclosed in their plasma membrane and their identical nuclei. Recall that the parent cell replicated its organelles before mitosis; thus, daughter cells contain a complete set of organelles smaller than the parent's cell size.

Cytokinesis in plant cells is different because plant cells have rigid cellulose cell walls which do not permit cytokinesis by furrowing. In plants, vesicles containing cellulose move to the middle of the cell. Additional vesicles arrive and coalesce, building a *cell plate* of cellulose (cell wall). When the cell plate is complete, the parent has divided into two separate daughter cells.

Nuclear membrane reorganization during cell division

Before mitosis begins, the cell's DNA is contained inside the nucleus, inaccessible to the mitotic spindle. During prophase and metaphase, the nucleolus disintegrates, chromatin condenses (i.e., heterochromatin), and the nuclear membrane breaks down.

This process exposes chromosomal DNA to the cell's cytoplasm and allows mitosis to divide the parental (original) cell into two identical daughter (new) cells.

The nuclear membrane does not reform until telophase, when the two daughter nuclei are each enclosed within nuclear membranes from fragments of the parental cell's nuclear membrane. Chromosomes then uncoil into relaxed chromatin (i.e., euchromatin), and the nucleoli reform, making nuclear reorganization complete.

Centrioles, asters and spindles

As discussed, centrioles are the microtubule-organizing centers of the cell. Centrioles replicate during interphase in preparation for mitosis.

During prophase, *polar microtubules* emerge from pairs of centrioles and centrosomes and push against each other and move the centrosomes to opposite sides of the cell.

Astral microtubules extend from the centrioles to assist in orienting the mitotic spindle apparatus.

At the end of prophase, kinetochore microtubules originate from the centrioles and attach to the chromosomes' kinetochores.

The spindle apparatus consists of two centrosomes (composed of two centrioles each), polar microtubules, astral microtubules (asters), and *kinetochore microtubules* (k-fibers) attached to the kinetochores on chromosomes.

Like animal cells, plant cells have a spindle apparatus, but many do not have centrioles or astral microtubules. Centrioles are not strictly necessary for mitosis, even in animal cells.

Chromatids, centromeres and kinetochores

As the cell enters metaphase, kinetochore microtubules extend from the centrosomes and attach to *kinetochores* on the chromosomes.

Kinetochores are assembled on the *centromere*, the chromosome region linking two sister chromatids.

There are two sections of the kinetochore: the inner kinetochore, which associates with the DNA of the centromere, and the outer kinetochore, which interacts with the kinetochore microtubules attached to the centriole.

During metaphase, microtubules pull on the chromosomes with tension, eventually aligning them at the metaphase plate. After successful attachment of chromosomes to the spindle via the kinetochore and microtubules, proteins are released from the kinetochore, which signals the end of metaphase and the beginning of anaphase.

Mechanisms of chromosome movement

In anaphase, the centromere holding the sister chromatids (S phase of interphase when the chromosome replicated to form two sister chromatids) dissolves, and the sister chromatids are released from their attachment point. The chromosomes are pulled to opposite sides of the cell by the shortening of kinetochore microtubules.

Shortening occurs when the motor protein attached to a kinetochore "walks" along the kinetochore-microtubule, dissembling the microtubule into tubulin subunits as it passes.

Polar microtubules assist the separation of chromosomes by lengthening the spindle. Wherever the ends of two polar microtubules from opposite poles overlap, motor proteins interact between the fibers and push them in opposite directions, thus pushing the entire spindle apart.

Phases of the cell cycle: G_0, G_1, S, G_2, M

Interphase is divided into three phases: G_1, S, and G_2.

G_0 is another resting phase when the cell is not dividing nor preparing to divide. G_0 is the static state in which cells remain permanently (e.g., nerve cells) and others temporarily.

A cell exits G_0 and reenter the active cell cycle upon receipt of signals from *growth factor* proteins.

Under a microscope, a cell is recognized as being in interphase by the lack of visible chromosomes because the DNA is uncoiled as loose chromatin (i.e., euchromatin).

During the *G_1 phase* (*Gap phase 1*), the cell continues normal function as it grows larger and replicates its organelles, including ribosomes, mitochondria, and chloroplasts (if a plant cell).

The mitochondria generate sufficient storage of energy for the functions of mitosis.

In G_1, the cell synthesizes mRNA and proteins in preparation for the *S phase*.

In the S (*Synthesis*) phase, the cell replicates DNA and produces two sister chromatids attached at the centromere. During the S phase, DNA nucleotide damage must be detected and fixed before the cell proceeds.

G_2 phase (*Gap phase 2*) is where the cell grows and synthesizes proteins needed for mitosis.

Completing G_2 marks the end of interphase and the beginning of mitosis, abbreviated as M.

Sequential phases of the cell cycle

The sequential phases of the cell cycle are:

G_0 = no DNA replication or cell division

G_1 = making of organelles, increase in cell size (growth)

S = DNA replication, complete duplication of chromosomes (sister chromatids)

G_2 = making of organelles, increase in cell size (growth), cell committed to mitosis

M = mitosis (PMAT); prophase, metaphase, anaphase, and telophase

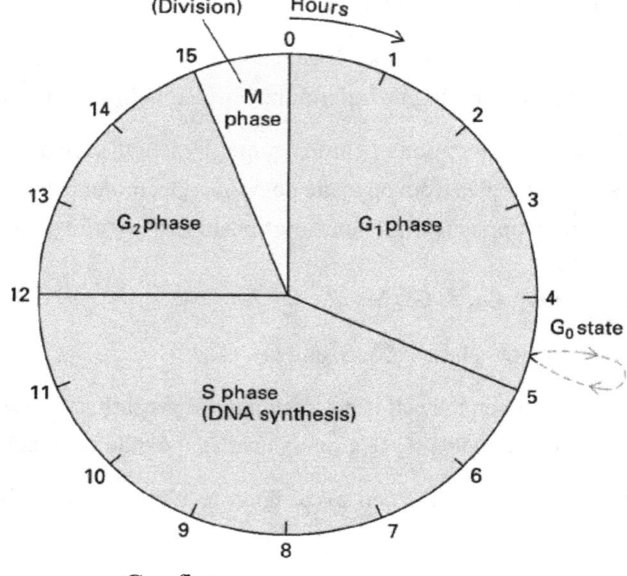

G₁ - first gap

S - DNA synthesis (replication)

G₂ - second gap

M - mitosis

The cell cycle divided between interphase and mitosis

Growth arrest

Cell growth is *cell proliferation* (for populations of cells) or an individual cell's growth, whereby biomolecules are synthesized.

Cell proliferation is the goal of unicellular organisms.

However, in multicellular organisms, cell proliferation must be carefully monitored to prevent tumor formation and invasion into nearby tissues.

Contact inhibition, the tendency for cells to cease dividing when they come into physical contact with their neighbors, regulates this. Therefore, lack of free space signals growth arrest.

The growth of the individual cells regulates cell populations.

For individual cell development, growth arrest is *cellular quiescence* or *cell cycle arrest.*

During the cell cycle, the cell encounters checkpoints where the cell cycle is halted before proceeding.

Checkpoints during cellular division

1. **The G_1 Checkpoint** – *Restriction Point*

    Partway through G_1, the cell reaches the restriction point.

    The cell checks for biomolecules, nutrients, and growth factors.

    If the cell is not sufficiently prepared for mitosis, the cycle halts, and the cell returns to G_0.

    Additionally, if DNA damage is detected, this triggers *apoptosis* (cell death) if the DNA is not repaired.

    If there are no inhibitory signals, the cell clears the checkpoint and proceeds toward DNA replication (S phase).

    G_1 is the checkpoint mediated by extracellular signals.

    After this point, intracellular signals direct the cell cycle to proceed or halt progress.

2. **The G_2 Checkpoint** – *DNA Damage Checkpoint*

    At the end of G_2, there is another checkpoint before the cell proceeds with mitosis.

    The cell checks for size and proper DNA replication.

    If the DNA has not finished replicating or damaged and requiring repair, it remains in G_2 until these issues are resolved.

3. **The M Checkpoint** – *Mitotic Spindle Checkpoint*

After the cell has entered mitosis and has reached metaphase, a final checkpoint occurs.

The M checkpoint ensures that the correct number of chromosomes are aligned at the mitotic plate and secured to the mitotic spindle.

Errors during chromosome segregation (e.g., nondisjunction) can cause defects resulting in genetic conditions (e.g., Down syndrome).

The M checkpoint reduces defects by arresting the cell in metaphase until the chromosomes are correctly attached to the spindle apparatus and aligned for anaphase.

Cell Cycle Control

Cyclin levels regulate cell division

The cell cycle is controlled by intracellular and extracellular signals that stimulate or inhibit metabolic events. Extracellular stimulatory signal molecules are growth factors; these are proteins or hormones that promote cell growth and differentiation.

Extracellular inhibitory signal molecules are growth suppressors or *tumor suppressors* because they prevent cancer cells' rampant growth.

Tumor suppressors inhibit growth by halting the cell cycle or directing the cell for apoptosis (i.e., cell death).

Intracellular signaling directs the cell cycle and involves the activation and inactivation of proteins; *cyclin-dependent kinases* (CDKs).

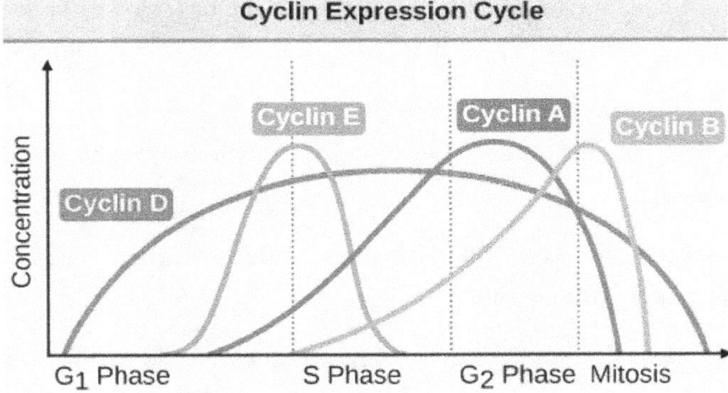

Relative concentrations of cyclin proteins during phases of the cell cycle

Apoptosis as programmed cell death

Apoptosis is the programmed cell death that occurs in multicellular organisms. It occurs in developing tissue (e.g., embryonic development where parts of tissues are no longer needed) and adult tissue (i.e., cell death balances cell division).

Apoptosis destroys cells that threaten the organism, such as infected cells, cells with DNA damage, cancerous cells, and immune system cells no longer needed and are unnecessarily attacking other body cells.

While paradoxical, apoptosis is essential for growth. For example, cells must die to create spaces in the webbed hands of a fetus for the formation of separate digits.

Some internal or external pathways and signals cause cell death, but the morphological changes during apoptosis are consistent whatever the cause.

These changes include shrinkage and *blebbing* (bulging) of the plasma membrane and nuclear envelope and DNA fragmentation.

Engulfment by nearby phagocytic cells occurs as a result.

Apoptotic cells release signals which attract phagocytic cells.

The engulfment of the dying cell's fragments prevents viruses or other dangerous cell contents from spilling out of the damaged cell.

Cancer cells lose cell cycle controls

Cancer cells are abnormal cells with various dangerous properties, making them a serious threat to the body. They invade and destroy normal tissue, causing serious illness and death.

Cancerous cells do not normally respond to the body's control mechanism. They no longer respond to inhibitory growth factors and do not require as many stimulatory growth factors.

Cancer cells may produce the required external growth factor (or override factors) or possess abnormal signal transduction sequences that falsely convey growth signals, thereby bypassing normal growth checks.

Due to their irregular growth cycles, if cancer cells' growth does occur, it does so at random points of the cell cycle.

Cancer can kill the organism because these cells can divide indefinitely (i.e., immortalized cells) if given a continual supply of nutrients.

DNA segments of *telomeres* form the ends of chromosomes and shorten with each cell division, eventually signaling the cell to stop dividing.

However, cancer cells produce telomerase (enzymes end with ~*ase*), which keeps telomeres long and allows cells to continue dividing as "*immortal.*"

Unlike normal cells, which differentiate, cancer cells are non-specialized.

Cancer cells do not exhibit contact inhibition; they do not avoid crowding neighboring cells but rather pile up and grow on one another. This behavior creates the characteristic tissue mass as a tumor.

Not all tumors are dangerous.

A *benign tumor* is encapsulated and does not invade adjacent tissue.

However, benign tumors can still compress and damage nearby tissue, and some benign tumors have the potential to become *malignant* (cancerous).

Malignancy occurs when new tumors are spread to areas distant from the primary tumor by *metastasis.*

Angiogenesis, oncogenes and tumor suppressor genes

Angiogenesis, the formation of new blood vessels, is a process required for metastasis. Angiogenesis is triggered when cancer cells release a growth factor that causes nearby blood vessels to grow and transport nutrients and oxygen to the tumor.

Because of angiogenesis in metastasis, angiogenesis inhibitors are an important class of cancer drugs.

Cancer cells have abnormal nuclei that may be enlarged and have an abnormal number of chromosomes, as some chromosomes are mutated, duplicated, or deleted.

When the DNA repair system fails to correct mutations during DNA replication, damage to crucial genes may occur.

Oncogenes encode growth factor proteins such as Ras, which stimulates the cell cycle (analogous to a *gas pedal* in a car).

In contrast, *tumor-suppressor genes* encode proteins such as p53 that inhibit the cell cycle (analogous to the *brake pedal* in a car).

Mutations of oncogenes or tumor suppressors can cause cancer.

Mutation of *proto-oncogenes* may convert them into *oncogenes*, which are cancer-causing genes.

An oncogene can cause cancer by coding for a faulty receptor in the stimulatory pathway, for an abnormal protein, or for abnormally high levels of a normal product that stimulates the cell cycle.

More than 100 oncogenes have been identified; the *ras* gene family includes variants associated with lung cancer, colon cancer, pancreatic cancer, leukemia, and thyroid cancers.

Mutation of tumor-suppressor genes results in unregulated cell growth.

For example, the *p53* tumor-suppressor gene is frequently mutated in human cancers than other known genes; it usually functions to trigger cell cycle inhibitors and stimulate apoptosis.

However, if it malfunctions due to mutation, cell growth is not suppressed, and cancer may result.

Editor's note: Nomenclature in molecular biology uses an italicized lower-case abbreviation for the gene (e.g., p53), while the protein (e.g., Ras) begins with a capital letter but is not italicized.

Biosignaling

Rhythmic fluctuations in the abundance and activity of cell-cycle control molecules pace the events of the cell cycle. One example of these control molecules are *kinases*, proteins which activate or deactivate other proteins by phosphorylation. They are responsible for a procession through the G_1 and G_2 checkpoints.

To give the signal, the kinases themselves must be activated by a *cyclin* protein. Because of this requirement, these kinases are called cyclin-dependent kinases (CDKs). Cyclins, named for their cycling concentration in the cell, accumulate during the G_1, S, and G_2 phases of the cell cycle.

By the G_2 checkpoint, enough cyclin is available to form a complex of cyclin and CDK called the *maturation-promoting factor* (MPF), or the M-Phase-promoting factor. MPF initiates progression from the G_2 to the M phase by phosphorylating key proteins involved in mitosis.

Later in mitosis, MPF switches itself off by initiating a process which leads to the destruction of cyclin. Cdk, the non-cyclin component of MPF, persists in the cell in an inactive form until it associates with the new cyclin molecules synthesized during interphase of the next round of the cell cycle.

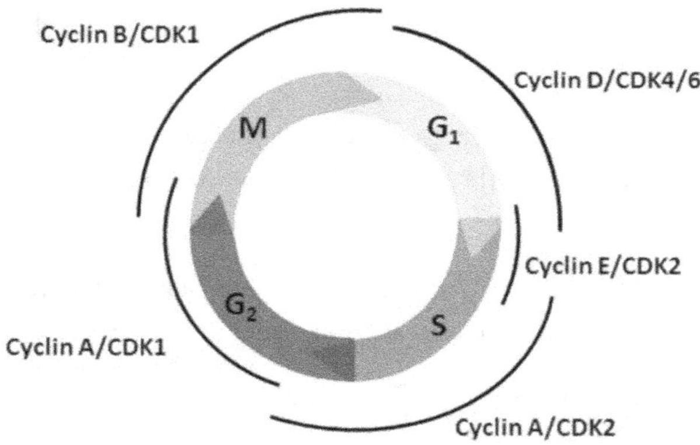

Cyclin-dependent kinases (CDK) and cyclin relationship to regulate the cell cycle

Platelet-derived growth factor (PDGF) is another protein that regulates cell growth and division. PDGF is required for the division of *fibroblasts*, connective tissue cells essential in wound healing. When an injury occurs, platelet blood cells release PDGF, which then binds to fibroblast receptors and activates a signal-transduction pathway that leads to a proliferation of fibroblasts and healing of the wound.

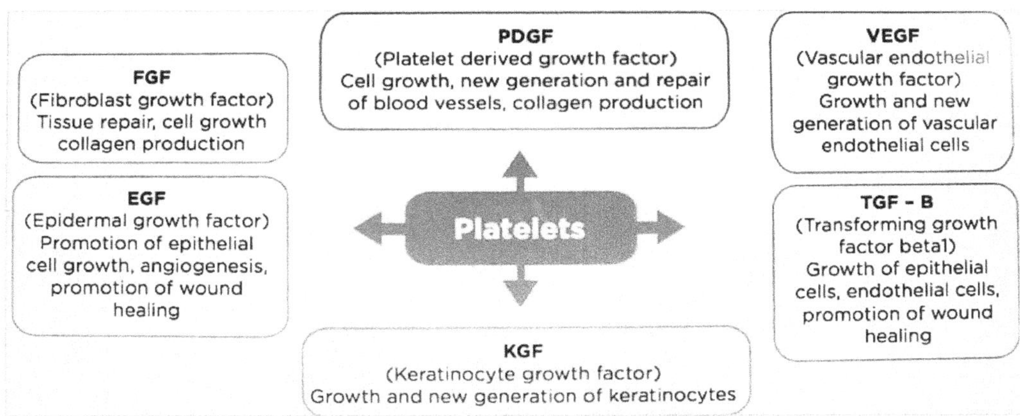

Growth factors released by platelets in wound healing

The extracellular environment has a direct effect on cell division. Cells grown in culture rapidly divide until a single layer of cells is spread over the area of the petri dish. However, if cells are removed, those bordering the open space begin dividing again and continue to do so until the gap is filled. This propensity to avoid division when in contact with neighboring cells is known as density-dependent inhibition of growth. When a cell population reaches a critical density, the amount of required growth factors and nutrients available to each cell becomes insufficient to allow continued cell growth.

Anchorage is another extracellular factor that controls cell division. For most animal cells to divide, they must be anchored to a substratum, such as the extracellular matrix of a tissue or the inside of a culture plate. Anchorage is signaled via pathways involving membrane proteins and the cytoskeleton.

Notes for active learning

Notes for active learning

Notes for active learning

REVIEW

Enzymes
&
Cellular Metabolism

Page intentionally left blank

Enzyme Structure and Function

Enzymes catalyze biological reactions

Enzymes assist most chemical reactions that occur in living organisms. Enzymes are biological molecules that act as catalysts by increasing the rate of chemical reactions. Most enzymes are proteins, although RNA molecules (ribozymes) can catalyze reactions. An enzyme cannot force a reaction to occur if it would not usually occur (i.e., products are less stable than reactants); it makes a reaction occur faster. Enzymes are highly specific in their action and catalyze a single reaction or class of reactions. Enzymes are not consumed in a reaction, so only small amounts of enzymes are needed in a cell. The overall 3D shape (tertiary structure) plays an essential role in the enzyme's function.

Enzymes are often named for their substrates by adding the suffix "*–ase*." For example, ribonuclease, abbreviated as RNase, is an enzyme that catalyzes the degradation of ribonucleic acid (RNA) into smaller components.

For example, the enzyme hexokinase is written above or below the reaction arrow. The reactants (starting material) upon which an enzyme acts, written to the left of the reaction arrow, are substrates. An enzyme may bind to one or more substrates (a molecule that enzymes act upon).

In the example below, there are two substrates: glucose and adenosine triphosphate (ATP). The products (to the right of the reaction arrow) are glucose-6-phosphate and adenosine diphosphate (ADP). Intermediates (not pictured in this example) are compounds temporarily formed between initial reactants and final products.

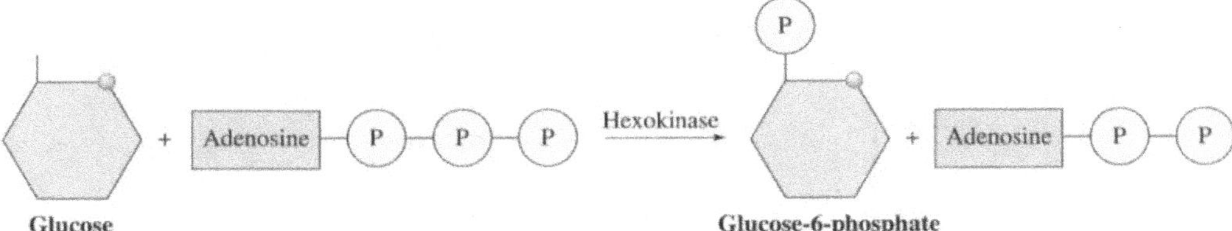

The substrates glucose and ATP are catalyzed by hexokinase to form glucose-6-phosphate and ADP. The hexokinase is not consumed in the reaction and continues the process with new substrates.

Reduction of activation energy

Reactants must reach a certain energy level before the reaction can convert the substrate(s) to the product(s). This energy is the *activation energy* (E_a) needed to break bonds within the reactants and thus enable them to react to form products. At a later stage in the reaction, energy is released as new bonds form within the product.

Enzymes decrease the activation energy of a reaction by lowering the energy of the transition state. The *transition state* (bond making and bond breaking state) is where the reactants are in an activated complex. As shown in the image below, the transition state corresponds to the highest energy level—the activation energy.

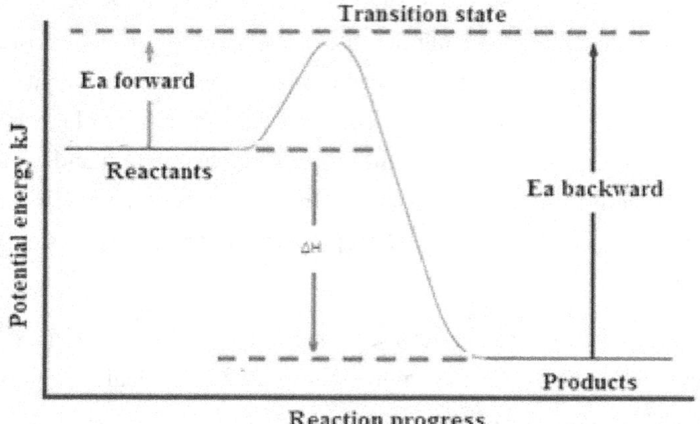

The energy is on the y-axis and time is shown on the x-axis. The products are lower in energy than reactants, and the reaction is spontaneous. The enzyme speeds the rate of spontaneous reactions.

Activation energy is an energy barrier to the reaction, which the reactants must overcome before they can be converted into products. This can be accomplished non-enzymatically by increasing the temperature, which increases molecular collisions between reactants. However, homeostasis means the body temperature is maintained within narrow ranges in many organisms. The same enzyme may lower the activation energy barrier for both the forward and reverse reaction, increasing both reaction rates, or another enzyme catalyzes each reaction separately.

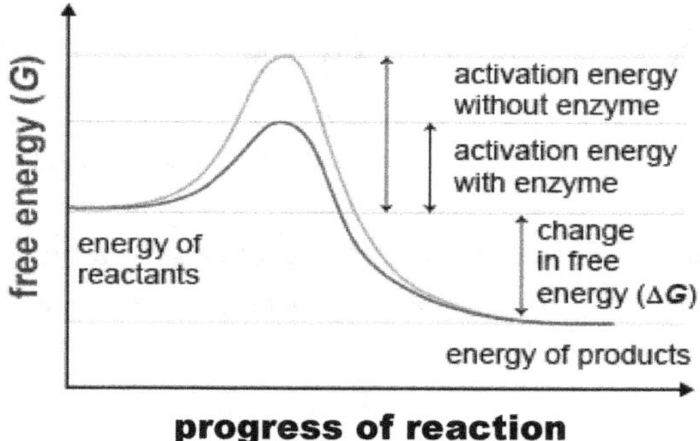

In the graph, the enzyme makes it easier for reactions to occur by lowering the required activation energy, but it does not alter the net energy release, ΔG (Gibbs free energy).

The enzyme provides an alternate pathway for reactants to form products. The lowering of the activation energy, which is accomplished during the formation of the enzyme-substrate complex, occurs in several ways:

Proximity: When the enzyme-substrate complex forms, the substrates are nearby, and therefore do not have to find each other as in a solution, thereby lowering the entropy (i.e., the disorder) of the reactants.

Optimizing orientation: The enzyme holds the substrates in the correct alignment and at the appropriate distance, usually by aligning active chemical groups. This lowers the entropy of the reactants.

Modifying bond energy: While binding to the substrate, the enzyme may stretch or distort a bond and weaken them so that less energy is needed to break the bond.

Electrostatic catalysis. Acidic or basic amino acids in the active site of the enzyme (a portion of the enzyme where the substrate binds) may form ionic bonds with the intermediate, which stabilize the transition state and lower the activation energy.

Substrates and enzyme specificity

When an enzyme binds to one or more substrates, it forms an enzyme-substrate complex, which is held together by hydrogen or ionic bonds. While bound, the catalytic action of the enzyme converts the substrate(s) to the product(s).

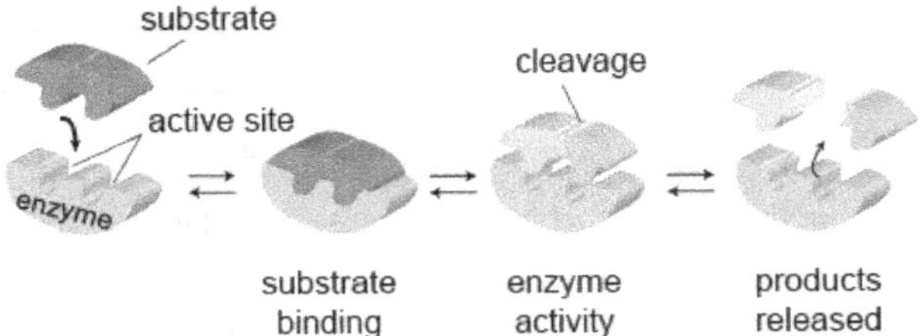

The substrate binds to active site of the enzyme with hydrogen or ionic bonds

An enzyme can distinguish its substrate from similar molecules and even isomers (same molecular formula but different molecules) of the same molecule. This enzyme-substrate *stereospecificity* refers to an enzyme's ability to distinguish between stereoisomers (different shapes) of the same molecule. For example, hexokinase binds D-glucose but not its stereoisomer L-glucose.

Enzyme classification by reaction type

While the names of some enzymes such as RNase describe their function, many enzymes have common names that do not refer to the substrate that they bind. Scientists created an enzyme classification system based on reaction type, with six main classes of enzymes. Note that within these six categories there are many subcategories. Knowing the reaction that a particular enzyme catalyzes is essential to categorization.

Oxidoreductases catalyze reactions that transfer electrons (oxidation-reduction or redox reactions). Common names for enzymes in this category are dehydrogenase, reductase, and oxidase.

Transferases catalyze group transfer reactions, whereby a group is transferred from a donor to an acceptor molecule. Hexokinase is a transferase that catalyzes the transfer of a phosphate group from ATP to glucose.

Hydrolases catalyze hydrolysis, in which water is used to break a bond (hydrolysis). Hydrolases transfer a functional group from a donor molecule to water (acceptor) and therefore are a specific type of transferase. An example is the digestive enzyme chymotrypsin, which can hydrolyze amide (amino acids) and ester (fatty acids) bonds.

Lyases catalyze the breakage of bonds using mechanisms other than hydrolysis (water) and oxidation (electron transfer). Lyases cleave double bonds by the addition of functional groups along with the reverse reaction (i.e., double bond formation via the removal of functional groups). Examples of lyases include fumarase (removes water), and adenosine deaminase (removes ammonia).

Isomerases produce isomeric forms (same atoms but the different connection between atoms) by transferring functional groups within a molecule, which allows for geometric or structural changes. For example, phosphoglucose isomerase converts glucose 6-phosphate to fructose 6-phosphate.

Ligases form covalent bonds by joining two molecules. A linkage reaction is usually coupled with an energy-producing reaction, such as the breakdown of ATP to ADP. An example is DNA ligase, which catalyzes the formation of a phosphodiester bond by joining DNA nucleotides together during replication of the genetic material.

Active site model

Every enzyme has an active site, which is a region on the surface of the enzyme where it binds to the substrate, orients it and facilitates the reaction. There are two prominent hypotheses on how substrates bind to the enzyme's active site. The *lock and key model* proposes that there is only one rigid active site that precisely fits the reactants, and the substrate fits inside its active site as a key fits into a lock. In this model, there is no modification of the enzyme after the substrate binds to the active site.

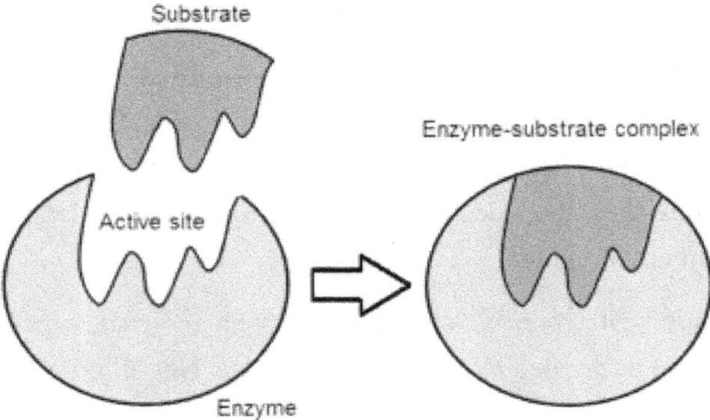

Lock and key model proposes that the enzyme has a shape matching the substrate and enzyme shape is static

Induced fit model

As an alternative to the lock and key model, the more recently developed *induced fit model* proposes that there is not a perfect match between the active site of an enzyme and the substrate that it binds. In this model, a small conformation (rotation) change occurs as the enzyme and substrate come together, which allows the active site to bind more precisely with the substrate.

The induced fit model hypothesizes that certain amino acid residues (i.e., side-chains) in the active site facilitate the enzyme finding the proper substrate. The initial interaction between the enzyme and substrate is weak, but these weak interactions cause conformational changes in the enzyme that strengthen binding. The energy barrier is lower in the "closed" form of the enzyme, where the active site fits snugly around the substrate.

Think of a handshake where the fingers move to more tightly grasp the other hand. The induced fit is the favored model of enzyme-substrate binding.

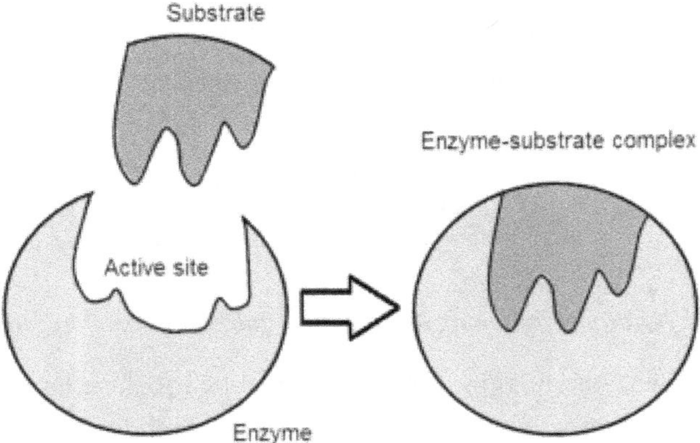

Induced-fit model proposes that conformational changes resulting from rotation around single bonds in the enzyme increase the interactions between enzyme and substrate.

Catalysts: cofactors, coenzymes, and water-soluble vitamins

When the enzyme and the substrate form the enzyme-substrate complex, the R-groups (i.e., side chains) of the amino acids in the active site catalyze the reaction. They often pull or contort the substrate, temporarily weakening bonds or altering the substrate's conformation. In reactions with two or more substrates, the side chains form a template to guide the substrates into the most energy-efficient conformation.

Some enzymes require *cofactors*, which are non-protein molecules that assist in chemical reactions that cannot be performed by the active site alone. Not all organisms can produce the cofactors needed, so they are required (essential) in the diet. Cofactors may be inorganic trace elements or metal ions, such as Cu^{2+}, Fe^{3+} or Zn^{2+} (electron carriers), or organic molecules as coenzymes.

Coenzymes are small, complex organic molecules, usually derived from vitamins, facilitating enzymatic reactions. Vitamins can be fat-soluble (A, D, E and K) or water-soluble (B and C), and it is the water-soluble vitamins that generally act as precursors to coenzymes. Vitamin deficiency may lead to the lack of a specific coenzyme and thus a lack of enzymatic action. Coenzymes may be loosely or tightly bound to the enzyme. Loosely-bound coenzymes are *cosubstrates*.

Some coenzymes are so tightly bound to their enzyme that they cannot be removed without denaturing the enzyme, and a coenzyme is a *prosthetic group*. Prosthetic groups may be attached to enzymes with covalent bonds. However, the division between loosely and tightly bound coenzymes is not always clear. For example, NAD^+ can be loosely bound in some enzymes but tightly bound in others.

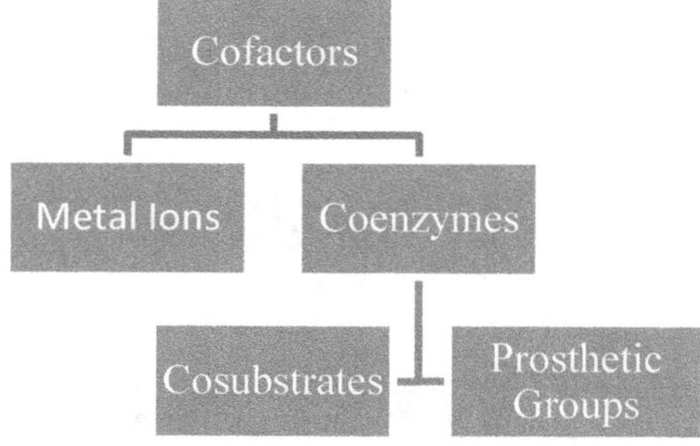

The relationship among molecules required by some enzymes

An *apoenzyme* is an enzyme missing the cofactor required to function, and the enzyme is inactive. A *holoenzyme* is an active form in which the cofactor and enzyme are bound together. Cofactors must be regenerated (i.e., returned to their original state) to complete the catalytic cycle. Cofactors, unlike enzymes, may change during the reaction and therefore need to be replenished.

Effects of local conditions on enzyme activity

Many factors affect enzyme activity, including substrate concentration, pH, temperature, and modulators.

The substrate in a cell is regulated by an organism's diet, absorption rates in the intestine, the permeability of the plasma membrane, or changes in intracellular breakdown and synthesis of the substrate. At a constant enzyme concentration, an increase in the substrate concentration increases the enzyme's activity by increasing the number of random collisions between the substrate and the active site on the enzyme. However, at some point, available active sites are bound, and increasing the substrate concentration has no further effect on enzyme activity, a state known as saturation.

Enzymes have an optimum pH at which they work most efficiently. The enzyme maintains its tertiary structure and, therefore, its active site at this pH. As the pH diverges from the optimum, enzyme activity decreases and may cease altogether.

Additionally, changes in pH may change the nature of an amino acid side chain. If an enzyme requires a carboxylate ion (COO^-), lowering the pH could convert the carboxylate ion to a carboxylic acid ($COOH$), which would cause enzyme activity to decrease.

An enzyme's pH optimum depends on the location of the enzyme. Enzymes in the stomach function at a much lower pH because of the stomach's acidic environment.

Enzyme	Location	Substrate	pH Optimum
Pepsin	Stomach	Peptide bonds	2
Sucrase	Small intestine	Sucrose	6.2
Urease	Liver	Urea	7.4
Hexokinase	All tissues	Glucose	7.5
Trypsin	Small intestine	Peptide bonds	8
Arginase	Liver	Arginine	9.7

Sometimes, the active site of an enzyme functions as a microenvironment more conducible to the reaction, such as providing a pocket of low pH in an otherwise neutral cell.

There is usually a temperature optimum at which an enzyme exhibits peak activity. Like the pH optimum, it is dependent on the environment in which the enzyme normally operates. For example, a DNA polymerase of a human has a lower temperature optimum than a DNA polymerase of a thermophilic bacterium. For most human enzymes, the temperature optimum is body temperature (37 °C).

As temperature increases, molecules move faster, with more random collisions between enzymes and substrates. Within the upper-temperature limit range, enzyme activity generally doubles with every 10-degree increase.

However, as with pH, at a certain point, the temperature gets too high, and bonds maintaining the 2°, 3° and 4° structure of the protein dissociate. This causes the enzyme's active site to become unstable, leading to a loss of function.

A *denatured* protein changes its three-dimensional shape. Denaturation alters the structure of an enzyme or another organic molecule so that the enzyme cannot carry out its intended function. The enzymes in bacteria are often denatured by high temperatures in processes like boiling contaminated drinking water and heat-sterilizing medical and scientific equipment.

However, enzyme activity is decreased by low temperatures, as when food is stored in a refrigerator or freezer. Enzymes are major participants in food spoilage, but low temperatures can significantly slow the spoilage process.

Modulators are compounds that modify the binding site of an enzyme, either inhibiting or enhancing the enzyme's activity. Modulators can have a strong effect on enzyme activity through either covalent (irreversible) or non-covalent (reversible) interactions between modulators and an enzyme.

Modulators are products of other chemical reactions or are activated by chemical signals. If modulators are required for enzymes to catalyze the reaction, they are cofactors. Modulators that increase an enzyme's catalytic activity are enzyme activators, enhancers or inducers; modulators that decrease or eliminate an enzyme's catalytic activity are enzyme inhibitors.

Control of Enzyme Activity

Kinetics: general (catalysis), Michaelis–Menten and cooperativity

Enzyme activity is studied using the principles of *enzyme kinetics*, which describe the rates of reactions and the effects of varying the conditions of a reaction. The *Michaelis-Menten kinetics* model of a single-substrate reaction is shown below.

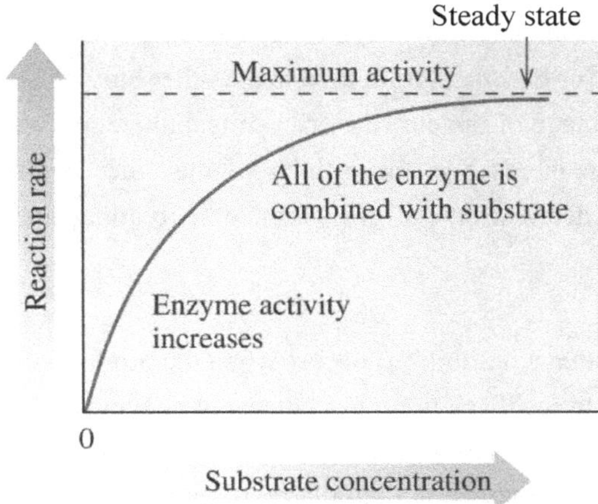

Enzymes become saturated when substrates occupy all of the active sites

The graph shows that the rate of enzyme-catalyzed reactions is not linearly correlated with substrate concentration. As the substrate concentration becomes higher, the enzyme becomes saturated with the substrate and approaches its maximum rate, V_{max}. K_m indicates an enzyme's affinity for a substrate. A low K_m reflects a high affinity for the substrate, while a high K_m reflects a low affinity. The Michaelis constant K_m is the substrate concentration at which the reaction rate is half of V_{max}.

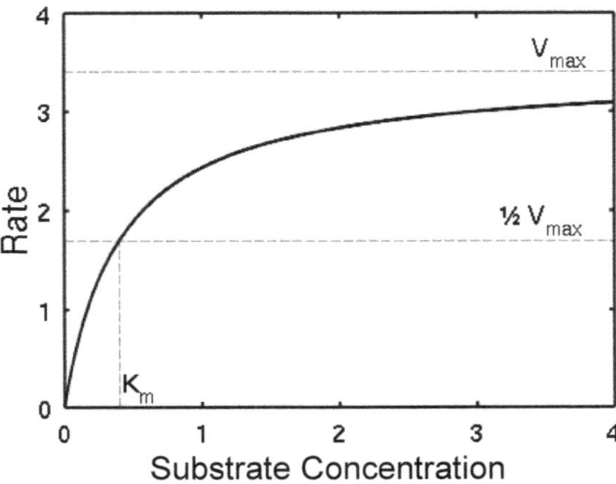

At V_{max} (maximum activity), the enzyme is operating in its optimum conditions, the steady-state. Under steady-state conditions, the substrate is efficiently converted to the product at maximum.

In the *reaction rate = k [A]·[B]*, the enzymes' rate depends on the constant *k,* which is a property of the enzyme. A higher *k* increases the speed of the reaction. Enzymes do not change ΔG, the net change in free energy between the products and reactants. Therefore, enzymes affect the kinetics (i.e., rate) of a reaction, but not the thermodynamics (relative stability).

Cooperativity can occur in enzymes with multiple binding sites. When one or more of these sites become activated (or deactivated), it affects the other binding sites on the enzyme, either by a conformational change of the enzyme or chemical alteration within the protein. Positive cooperativity causes an increase in the affinity of the other binding sites for a substrate. Negative cooperativity decreases the affinity of the other binding sites for a substrate.

Feedback regulation

Negative feedback (feedback inhibition) occurs when the product of a reaction deactivates an enzyme in the process. In feedback inhibition, the enzyme typically has one or more *allosteric sites* (regions other than the active site), where the reaction product can bind. When the reaction product binds, the enzyme is spatially rearranged or chemically modified and can no longer bind to the original substrate, halting the reaction. This can be done to prevent overproduction of a product or avoid consuming too many molecules that are energy sources for the cell.

For example, hexokinase, the first enzyme in glycolysis (the first pathway in cell respiration), is inhibited by its product, glucose-6-phosphate. This prevents the glucose substrate from being depleted. It is an example of negative feedback because the product is used as a signal to decrease the further output of the product.

Positive feedback (feedforward) is when a product is used as a signal to increase the further output of the product. An example is blood clotting, which is a catalytic cascade process. The cascade begins with an initial factor, such as catalyzing proteases, and additional steps follow, each step amplifying the preceding. Positive feedback is much less common than negative feedback.

Irreversible inhibitors

Inhibitors are molecules that reduce enzyme function. There are several types of behaviors that inhibitors may display. Some inhibitors are irreversible; they cause the enzyme to lose activity permanently because they become covalently bound to the enzyme and cannot be easily removed.

An irreversible inhibitor covalently bonds with amino acid side chains in the active site of the enzyme, excluding the substrate and thus blocking the catalytic reaction. Since blocking the activity of a particular enzyme can deactivate a pathogen or correct a metabolic imbalance, many medications are enzyme inhibitors. Penicillin is an example of an irreversible inhibitor of medical applications. Penicillin binds to the active site of an enzyme that bacteria use to synthesize cell walls. When the bacterial enzyme covalently bonds with penicillin, the enzyme loses its catalytic activity, and the growth of the bacterial cell wall slows. Without a proper cell wall for protection, the bacteria cannot survive, and the bacterial infection is alleviated.

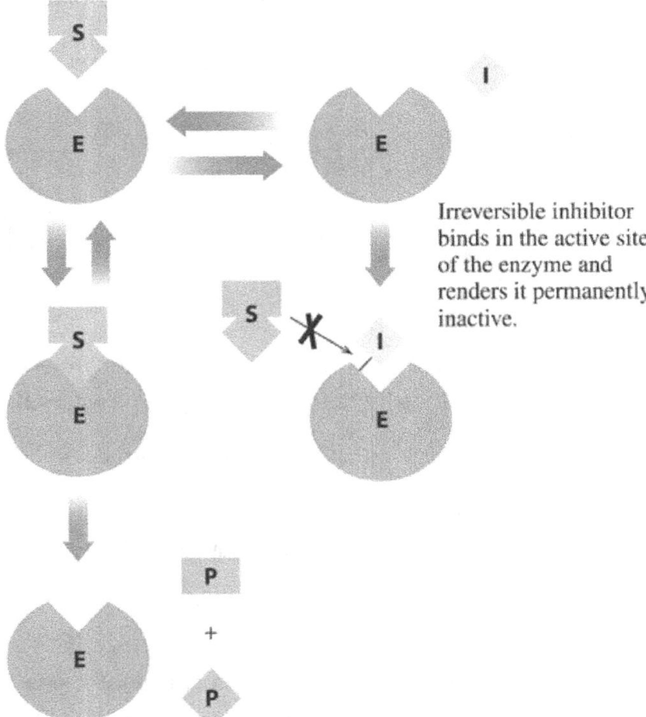

Irreversible inhibitors often bind to the active site with covalent bonds

Other inhibitors are reversible, which use hydrogen bonds, hydrophobic interactions, or ionic bonds and temporarily cause the enzyme to lose activity. In reversible inhibition, the inhibitor causes the enzyme to lose catalytic activity, but if the inhibitor is removed, the enzyme becomes functional again.

Inhibition types: competitive, uncompetitive, non-competitive, and mixed

There are four types of reversible inhibitors: competitive, uncompetitive, non-competitive, and mixed.

Competitive inhibitors are structurally similar to the substrate and bind to the enzyme's active site. The inhibitor directly competes with the substrate for the active site. When a competitive inhibitor binds to the active site, it prevents the substrate from binding to the same site. Only when the inhibitor has been released from the active site can the substrate bind.

The substrate binds to the enzyme at the active site to form a product

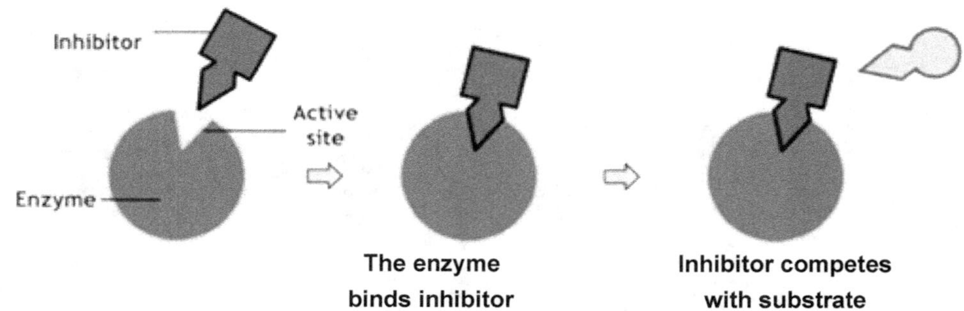

Competitive inhibition where the inhibitor binds to enzyme's active site

Competitive inhibitors usually bind directly to the active site. In other examples, they competitively may bind close to the active site in a way that blocks the active site. The definition of competitive inhibition is that the binding of the competitive inhibitor and binding of the substrate are mutually exclusive.

Malonate is an example of a competitive inhibitor. It is structurally similar to succinate, a substrate in the Krebs cycle of aerobic respiration which binds to the active site of a dehydrogenase enzyme. Malonate can compete with succinate for the active site and prevent succinate from binding, thereby inhibiting oxidation.

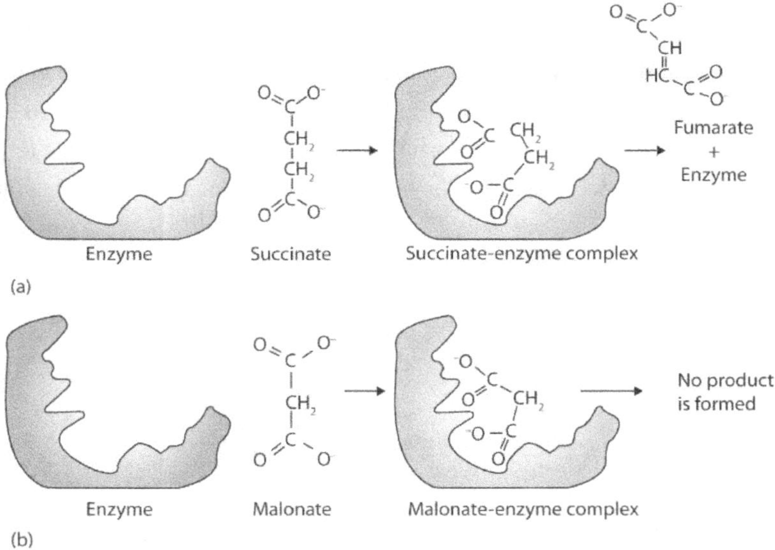

Competitive inhibition: a molecule shaped similarly to the substrate binds to the active site

The effects of competitive inhibition can be *reduced* by increasing the substrate concentration. The substrate can then outcompete the inhibitor for binding to the active site. As in the graph below, it is possible to reach the maximum rate of reaction V_{max} (or a rate close) in the presence of a competitive inhibitor. However, note that K_m increases because more substrate is needed to reach K_m (substrate concentration equal to $\frac{1}{2}V_{max}$), indicating lowered binding affinity of the substrate.

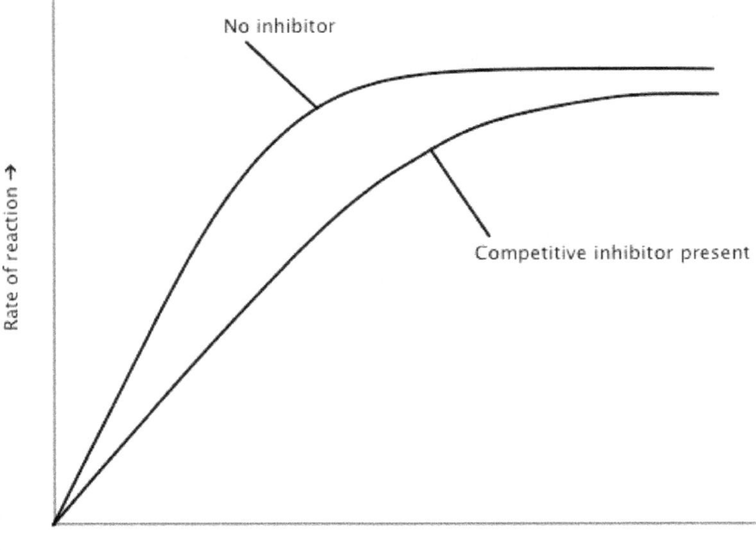

The effect of increasing substrate concentration on a competitive inhibitor where substrate outcompetes the competitor for enzyme binding

Uncompetitive inhibitors bind to the enzyme-substrate complex and render it ineffective. The uncompetitive inhibitor can only bind once the enzyme-substrate complex is fully formed. This type of inhibition is most effective when the substrate concentration is high. An uncompetitive inhibitor does not have to resemble the substrate of the reaction that it is inhibiting. Uncompetitive inhibition decreases V_{max} because it takes longer for the substrate or product to leave the active site, which decreases K_m. The decrease in K_m indicates a higher binding affinity of the substrate to the enzyme.

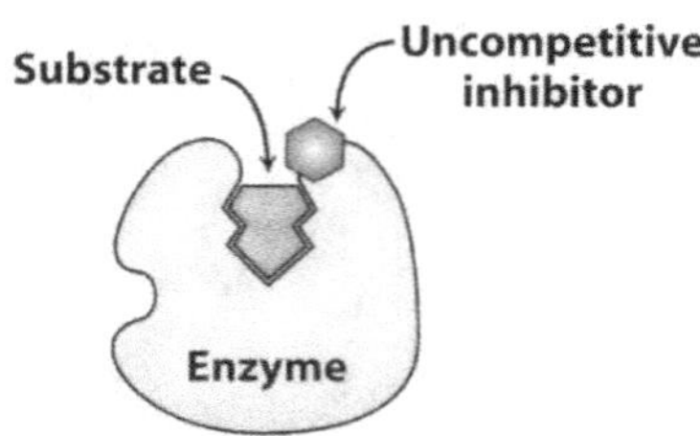

The uncompetitive inhibitor binds after the enzyme-substrate complex is formed

Non-competitive inhibitors are not structurally similar to the substrate, and they bind to an allosteric site (not the active site) of the enzyme. This binding changes the conformation (shape) of the enzyme, thereby reducing the affinity of the substrate to the active site. For this inhibition, the substrate may still be able to bind to the active site; however, the enzyme is not able to catalyze the reaction or does so at a slower rate.

Non-competitive inhibitors bind equally well to either the enzyme or enzyme-substrate complex, with no preference for either state. Increasing the substrate concentration cannot prevent a non-competitive inhibitor from binding to the enzyme. Therefore, some enzymes are inhibited when a non-competitive inhibitor is present, regardless of substrate concentration.

An example of a non-competitive inhibitor involves ATP. When ATP accumulates within the cell, it binds to an allosteric site on the phosphofructokinase-1 enzyme. When bound, it changes the enzyme conformation and lowers the rate of reaction so that less ATP is produced. This is an example of negative feedback inhibition.

The maximum rate of reaction V_{max} is lower in the presence of a non-competitive inhibitor, while the K_m remains unchanged because the substrate concentration to achieve $\frac{1}{2} V_{max}$ of both reactions is the same.

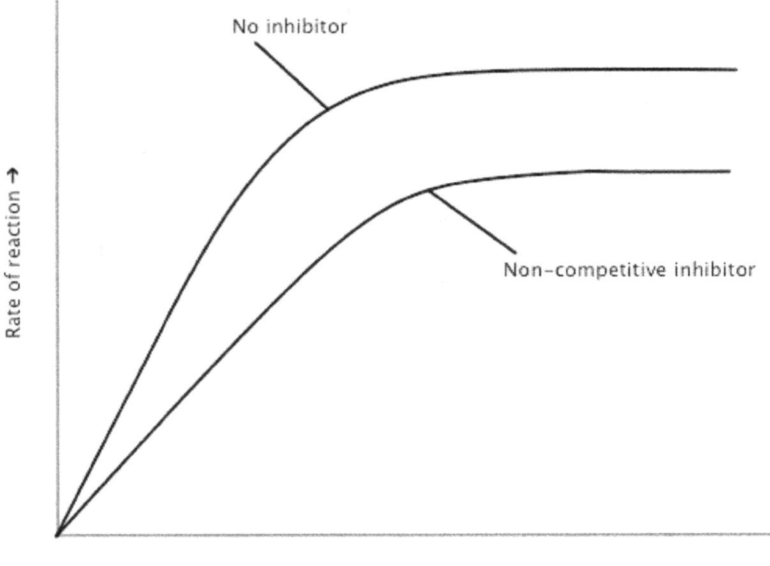

A non-competitive inhibitor binds to an allosteric site,
increasing substrate concentration cannot restore V_{max}

Mixed inhibitors, like non-competitive inhibitors, can bind to the enzyme at the same time as the substrate, but the binding of the substrate affects the binding of the inhibitor, and vice versa. Although the inhibitor can bind to the enzyme whether or not the substrate is bound, the inhibitor has a greater affinity for one of the states. Mixed inhibition affects the rate between that of competitive and uncompetitive inhibition.

Mixed inhibition is often grouped with non-competitive inhibition because both can bind to either the enzyme or enzyme-substrate complex. However, non-competitive inhibitors have an equal preference for either state, while mixed inhibitors prefer one state. Furthermore, while non-competitive inhibition is allosteric, it is possible for mixed inhibitors to bind at the active site, although they usually bind allosterically.

Increasing the concentration of the substrate can reduce the rate for mixed inhibition but cannot entirely overcome it. Mixed inhibition results in a decreased V_{max}, and can result in either an increase or decrease in K_m, depending if the inhibitor binds to the free enzyme or the enzyme-substrate complex.

Inhibition Type

	Competitive	Uncompetitive	Non-competitive	Mixed
Molecule(s) which bind	Enzyme	Enzyme-substrate complex	Enzyme-substrate complex or enzyme	Enzyme-substrate complex or enzyme (but prefers one or the other)
V_{max} effect	Unchanged	Decrease	Decrease	Decrease
K_m effect	Increase	Decrease	Unchanged	Increase or decrease
Mechanism of inhibition	Blocks substrate binding	Blocks substrate and enzyme from forming a product	Allosterically changes enzyme conformation	Allosterically blocks substrate binding or changes enzyme conformation

Allosteric and covalently-modified regulatory enzymes

There is a vast array of enzymes in organisms, but only regulatory enzymes undergo inhibition or activation. The initial mechanism by which enzymes are regulated is the expression of enzyme-encoding genes.

However, enzymes can be part of the regulatory mechanism. In cells, groups of enzymes can work together for metabolic processes; such as an *enzyme system*. The enzymes often act in successive pathways, where the product of one enzyme becomes the substrate of the next enzyme in the sequence. A sequence of enzyme-mediated reactions generally contains a rate-limiting reaction (the slowest step) and regulates the rate of the entire pathway.

Regulatory enzymes display increased (or decreased) activity in response to different signals. This is ideal for the production of molecules that may be needed in different amounts at different times (e.g., hormones). The most efficient position for the regulatory enzyme is the first enzyme in the pathway so that the intermediate products of the pathway are not unnecessarily synthesized, which wastes biomolecules and cellular energy.

There are two classes of regulatory enzymes in metabolic pathways: allosteric enzymes and covalently-modified enzymes. However, it is possible for an enzyme to use both types of regulation.

Allosteric enzymes, which are usually larger and more complex than simple non-regulatory enzymes, change their conformation when a *modulator* (effector) non-covalently binds to an allosteric site.

The binding of the effector causes a change in the enzyme's conformation, which leads to a reversible change in the binding affinity at the enzyme's active site. This is an example of cooperativity.

The allosteric effector can be either an activator that increases the enzyme's affinity for its substrates (positive cooperativity, positive allosteric modulation or allosteric activation) or an inhibitor that decreases the enzyme's affinity for its substrates (negative cooperativity, negative allosteric modulation or allosteric inhibition).

Due to the complex cooperative interactions between protein subunits, allosteric enzymes usually do not follow the typical hyperbolic Michaelis-Menten behavior of other enzymes. Instead, they display sigmoid kinetic behavior.

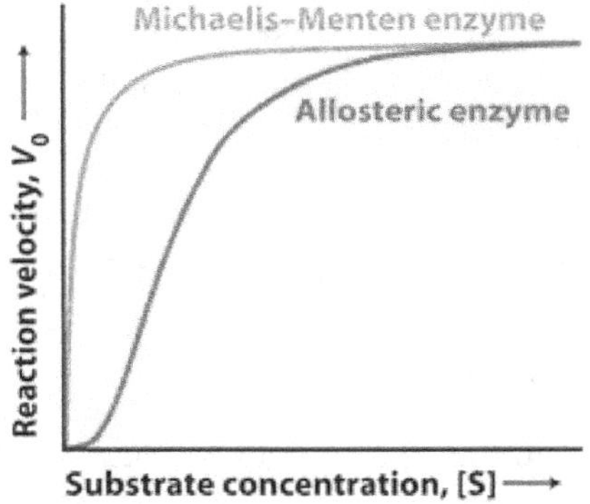

Allosteric enzymes show a sigmoidal (S-shaped) profile on a velocity vs. substrate concentration graph. They are proteins with quaternary structure involving two or more polypeptide chains (e.g., hemoglobin).

One example of allostery is when the end product of a pathway binds at an allosteric site on the first enzyme of the pathway, shutting down the pathway. This is negative feedback (feedback inhibition); more specifically, it is *end-product inhibition.*

When there is an excess of the end product, the whole metabolic pathway ceases as the end product inhibits the first enzyme of the pathway. This decreases the formation of intermediates and the end product.

When the levels of the end product decrease to a particular level, the end products that are bound to allosteric sites on the enzyme are released, and the enzymes become active again, turning on the metabolic pathway. In this example, the allosteric effector is an inhibitor.

Allostery may be observed in competitive, non-competitive, uncompetitive or mixed inhibition.

Allostery and the types of inhibition are often confused with one another. The types of inhibition are broad terms which describe how the inhibitor affects the V_{max} and K_m of an enzyme.

Allostery is a specific way that inhibition (or activation) can be accomplished, by binding at a site other than the active site, which causes a conformational change and alters the shape of the active site and the enzyme's activity. Inhibitors may exhibit allosteric behavior, but not all allosteric behavior is necessarily a part of inhibition or any enzyme regulation.

There are two classic models of allosteric regulation. Both of these hypothesize that enzyme subunits exist in either a relaxed state (higher affinity for a substrate) or a tensed state, (lower affinity for a substrate).

The *concerted model*, (symmetry model) states that all subunits in an enzyme must exist in the same conformation, either relaxed or tense, and the binding of an allosteric effector shift the conformation of the various sites from relaxed to tense, or vice versa.

The *sequential model* states that subunits are not necessarily all in the same conformation, and when an effector binds to an allosteric site, it does not propagate a conformational change to the other subunits; rather, it slightly alters the structure of the active binding sites so that they are either more or less receptive to a substrate.

Homotropic allosteric modulators are the substrates for the enzymes that they bind to. When a homotropic allosteric modulator binds to one active site of the enzyme, it alters binding affinity or catalytic activity at other active sites. Homotropic allosteric modulators are usually activators and an example of positive feedback.

Heterotropic allosteric modulators are more common, whereby most allosteric modulators are not both a substrate and modulator of the same enzyme. They can be either activators or inhibitors.

Allosteric enzymes are not the only regulatory enzymes that the body utilizes. Enzymes can be regulated by the addition or removal of chemical groups that are covalently bound to the enzyme. This changes the chemical properties of the active site, thus changing the enzyme's catalytic activity. The covalent addition of chemical groups activates some enzymes and inactivates others.

Many chemical groups can function as covalent modifiers, including phosphoryl, methyl, and uridyl. The covalent modification can be either reversible or irreversible.

A common covalent modification is phosphorylation (adding a phosphate group) or dephosphorylation (removing a phosphate group) of an enzyme's amino acid side chains because these are relatively easy biochemical processes.

Kinases catalyze the addition of a phosphate group (i.e., phosphorylation), while *phosphatases* catalyze the removal of a phosphate group (i.e., dephosphorylation).

Phosphorylation is important in regulatory pathways; an attached phosphate group gives an additional negative charge to the protein, which can be a source of electrostatic interactions that change the enzyme's conformation. This changes its capacity to form hydrogen bonds and alters its affinity for a substrate.

Phosphorylation can greatly amplify a signal, because one kinase may create an exponential effect, phosphorylation cascade.

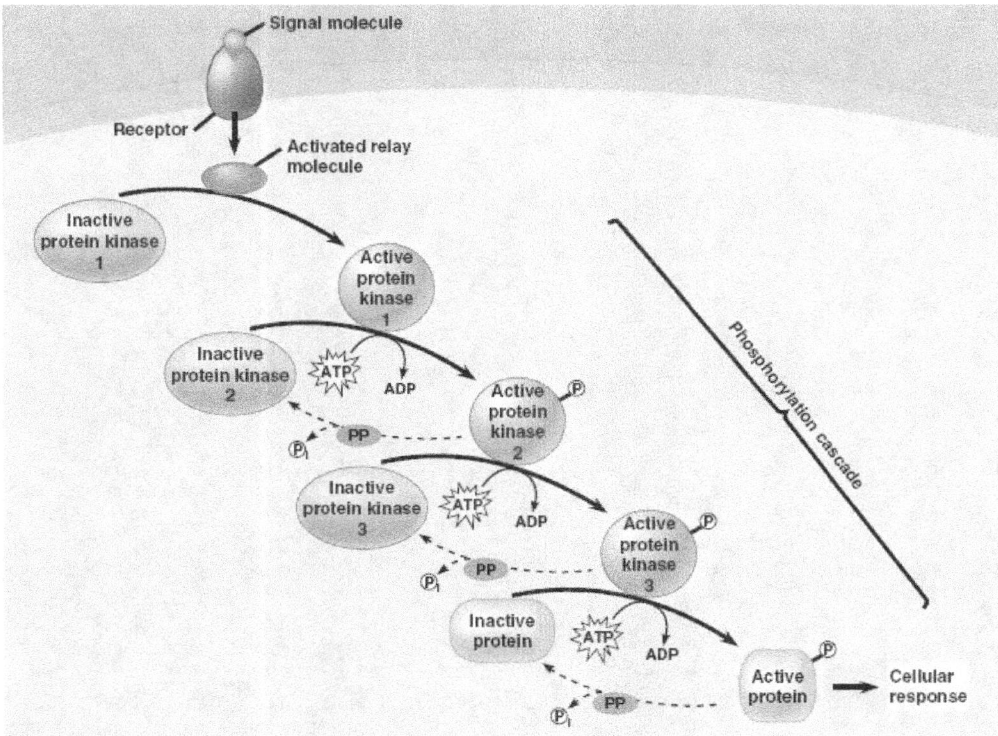

Proteolysis is another common covalent modification, by which specific peptide bonds in a protein are cleaved in a hydrolysis reaction. Proteolysis is often accomplished by protease enzymes but can be affected by other enzymes or non-enzymatic methods.

Zymogens

Zymogen (a proenzyme) is an enzyme that must undergo proteolysis to be activated. To form the final active enzyme, the zymogen requires a biochemical change such as proteolysis.

For example, the proteins involved in blood clotting circulate in the bloodstream as zymogens. Tissue damage or trauma activates enzymes that cleave the zymogens to make them active enzymes, which then cleave other enzymes, creating a cascade of clotting factor formation.

The last step is the cleavage of the protein fibrinogen into fibrin, which causes blood to clot. Proteolysis is irreversible, so if the enzyme has to be deactivated, enzyme inhibitors are required.

Notes for active learning

Cell Metabolism

Metabolism for energy production

Energy is the capacity to do work. Cells continually use energy to develop, grow, reproduce, and perform biochemical functions.

Kinetic energy is the energy of motion. Examples of kinetic energy include the beating of cilia, a contracting muscle, and the pumping of ions across a plasma membrane.

Potential energy is stored energy. The food contains *chemical energy* within bonds as potential energy.

Chemical energy can be transformed into other forms of energy through chemical reactions. For example, when a living organism digests food through a series of chemical reactions, the energy it obtains from the food moves its body; chemical energy is converted to kinetic energy.

Metabolism is the sum of the biochemical reactions in a cell. Metabolism consists of *anabolism* (synthesis) and *catabolism* (breakdown) of organic molecules required for cell structure and function.

Energy obtained from catabolism drives anabolism. The *metabolic pool* is molecules of biosynthesis.

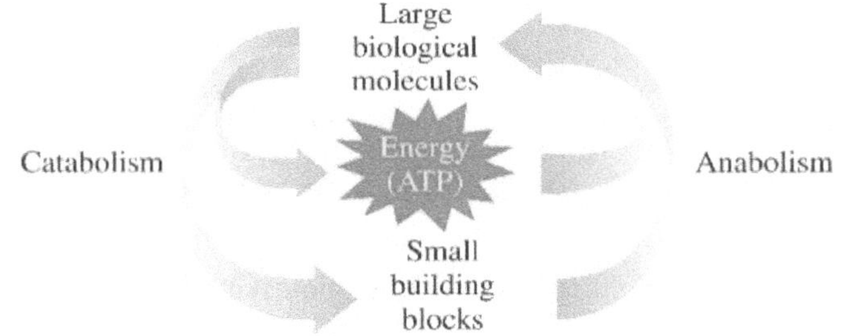

Metabolism includes catabolism (breakdown) and anabolism (synthesis)

Metabolic pathways

A *metabolic pathway* is an orderly sequence of linked reactions; a specific enzyme catalyzes each step in the pathway. Metabolic pathways begin with a *reactant* (a substance that participates in a reaction) and end with a *product* (a substance formed by the reaction). A and B are reactants in a reaction A + B → C + D, and C and D are products. Since pathways often use the same molecules, one pathway can lead to several others.

Metabolic pathways are compartmentalized into sections of the cell. Within organisms, energy must be transferred in small amounts to minimize the heat released in the process.

Reactions producing energy (i.e., exergonic) are coupled with reactions requiring energy (i.e., endergonic), thereby helping thermoregulation maintain constant body temperature for homeostasis.

Thermodynamics studies energy transformation and is an important concept in chemical reactions. Chemical reactions involve: 1) the breaking of chemical bonds in the reactants, which requires energy, and 2) the making of new chemical bonds to form products, releasing energy in the form of heat.

There are four laws of thermodynamics.

The *zeroth law of thermodynamics* states that if two systems are in thermal equilibrium with a third system, then they are in thermal equilibrium with each other. This explains the concept of temperature. Another way to understand this law is that two objects placed in contact with each other cause heat energy to transfer from one to the other until they reach thermal equilibrium.

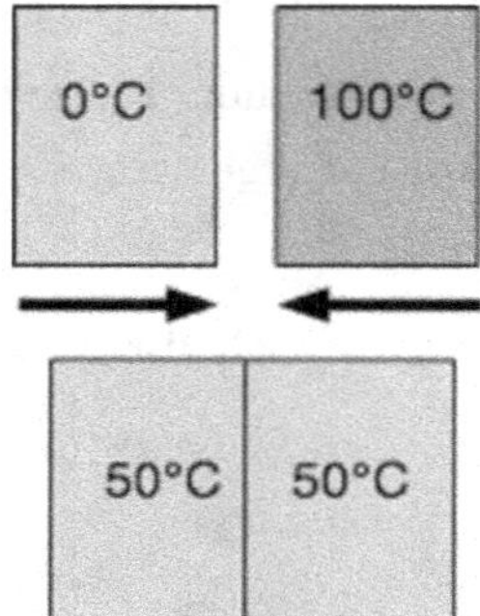

Zeroth law states that objects in contact transfer energy until equal

The *first law of thermodynamics* states that energy can be transferred and transformed, but it can neither be created nor destroyed. The total energy of the universe is constant. In an ecosystem, energy from sunlight is converted by photosynthesis to chemical energy in the form of sugars (e.g., glucose).

When an animal consumes the plant, some of the chemical energy in the plant is converted to chemical energy in the animal, which can eventually become kinetic energy or heat loss.

The *second law of thermodynamics* states that energy transfer or transformation increases the entropy (disorder) of the universe (i.e., the trend toward randomness). The tendency to increase disorder is the driving force in spontaneous reactions. To regain order, energy must be invested, which usually releases heat, increasing the entropy elsewhere.

The *third law of thermodynamics* states that the entropy of a system approaches zero as the temperature approaches absolute zero (0 Kelvin). Absolute zero is a theoretical value, and no system has been observed at absolute zero, where the motion of electrons would cease. When the system has a temperature above 0 K, particles start to move, and disorder is generated.

ΔG and K_{eq}: relationship of ΔG and equilibrium constant

Gibbs free energy (G) is a thermodynamic quantity which measures the useful work that a system can do. If the change in Gibbs free energy ΔG for a chemical reaction at a specific temperature T is known, the equilibrium constant (K_{eq}) can be calculated for the process using the equation:

$$\Delta G = -RT \ln(K_{eq})$$

where $R = 8.314$ J mol^{-1} K^{-1} and T is the temperature in Kelvin.

The equilibrium constant K_{eq} measures the extent to which the reactants are converted to products. When defining the K_{eq} for a chemical reaction, the concentrations of the products are in the numerator, and the concentrations of the reactants are in the denominator. The concentrations are raised to the power of the stoichiometric coefficient for the balanced reaction.

For example, for the balanced reaction aA + bB $\rightleftharpoons$ cC + dD, the K_{eq} is:

$$K_{eq} = [C]^c[D]^d / [A]^a[B]^b$$

Under specified conditions (i.e., same temperature and pH), the K_{eq} is the same for a given reaction. A higher K_{eq} means that a reaction is more likely to occur, while a lower K_{eq} means that a reaction is less likely to occur.

$K_{eq} < 1$: reactants are predominant

$K_{eq} = 0$: both reactants and products are equally favored

$K_{eq} > 1$: products are predominant

Concentration and Le Châtelier's Principle

Le Châtelier's Principle, (Equilibrium Law), describes equilibrium changes when the conditions of a reaction are altered.

When there is a change in pressure, volume, temperature or concentration, the position of equilibrium moves in an attempt to counteract that change.

For the following reaction in equilibrium (i.e., the rate of the forward and reverse reaction are equal):

$$A + B \rightleftharpoons C + D$$

If the concentration of A is increased, the position of equilibrium shifts to counteract the change. Therefore, the position of equilibrium moves toward the products, which increases the concentration of C and D and decreases the concentration of A.

If the concentration of A is decreased, the position of equilibrium moves toward the reactants, which increases the concentration of A and decrease the concentration of C and D.

Endothermic and exothermic reactions

Reactions can be classified as endothermic or exothermic, which refer to transfer of heat or changes in enthalpy (ΔH). *Enthalpy* is the heat transferred during a constant pressure process. In an *endothermic* reaction, ΔH is positive, and the reaction absorbs heat, which causes a decrease in the temperature of the surroundings.

Examples of endothermic reactions include the melting of ice, cooking an egg and dissolving ammonium chloride in water. All these processes require heat (i.e., it is easier to dissolve a substance in a hot solution).

In an *exothermic* reaction, ΔH is negative, and the reaction gives off heat, which causes an increase in the temperature of the surroundings. Examples of exothermic reactions include freezing water, burning propane and splitting an atom. These processes release heat.

Gibbs free energy: *G*

Free energy, or Gibbs free energy (G), is a thermodynamic quantity that represents the amount of energy that is available to do work after a chemical reaction. It is called "free" energy because this is the energy which can perform work, not because there is no energy cost to the system. The change in free energy ΔG refers to the difference between the free energy (relative stability) of the products and the reactants.

It is defined as $\Delta G = \Delta H - T\Delta S$, where ΔH is the change in enthalpy in joules, T is the temperature in Kelvin, and ΔS is the change in entropy. ΔG cannot be measured directly, but ΔH and ΔS can be determined, which allows ΔG to be calculated at a known temperature.

Spontaneous reactions and ΔG

Reactions can be classified as endergonic or exergonic. An *endergonic* reaction has a positive ΔG and requires energy to proceed. Since endergonic reactions require energy, these reactions are not spontaneous. Anabolic reactions, where metabolites combine to form larger molecules, tend to be endergonic ($+\Delta G$).

Exergonic reactions have a negative ΔG and result in a release of energy. These reactions are spontaneous. However, the spontaneous reaction does not necessarily occur rapidly. Catabolic reactions, where larger molecules are broken into smaller metabolites, tend to be exergonic ($-\Delta G$).

Phosphoryl group transfers and ATP hydrolysis

Adenosine triphosphate (ATP) is the energy currency of life. ATP is a nucleotide composed of the base adenine, the 5-carbon sugar ribose, and three phosphate groups. It is a relative of adenine, a nucleotide found in DNA.

Adenosine "triphosphate" derives its name from the three phosphates attached to the ribose of the molecule. The phosphate groups are linked by phosphoanhydride bonds, which have a large $-\Delta G$ on hydrolysis, and are therefore "high-energy" bonds.

ATP is an unstable molecule because the three phosphates in ATP are negatively charged and repel each other. Compounds with high-energy bonds have high group transfer potential, as chemical groups can easily be removed and transferred to another molecule.

Because of the high group transfer potential, the terminal phosphate group on ATP can be spontaneously removed. When the terminal phosphate group is removed via hydrolysis (addition of a water molecule), adenosine diphosphate (ADP), a more stable molecule results.

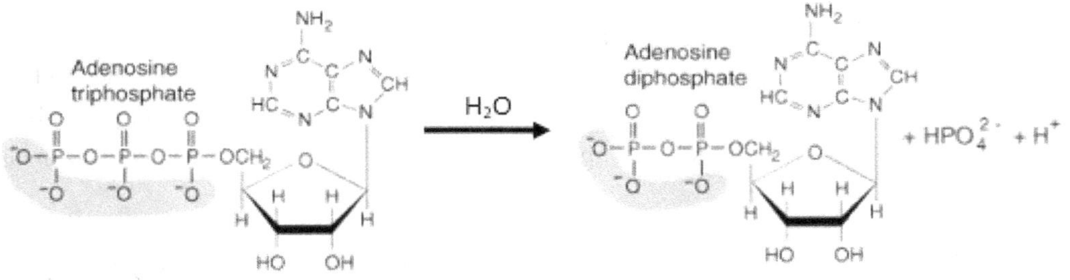

The change from a less stable to a more stable ADP molecule releases energy ($-\Delta G$)

The conversion of ATP to ADP is an exergonic reaction that releases about 7.3 kcal of energy per mole; this energy can be used for endergonic reactions in the cell that require energy (usually anabolic or synthesis reactions). ATP breakdown is coupled with endergonic reactions to minimize energy loss.

For cells to function, ATP must rapidly be regenerated. One muscle cell can consume and regenerate over 10,000,000 ATP per second. If ATP could not be regenerated, humans would have to consume nearly their body weight in ATP each day.

ATP breakdown is coupled with cell reactions that require energy. A *coupled reaction* is when energy released by an exergonic reaction is used to drive an endergonic reaction. The endergonic process absorbs the free energy released from the exergonic process.

It is possible for both of the high-energy phosphoanhydride linkages in ATP to be cleaved by hydrolysis. When two phosphate groups are removed, the resulting molecule is adenosine monophosphate (AMP), which serves as an energy sensor and regulator of metabolism. When there is a deficiency in the amount of ATP being produced, a higher proportion of the cell's adenine is in the form of AMP, and AMP stimulates the pathways that generate more ATP.

The synthesis of ATP involves endergonic reactions that require energy. The pathways of cellular respiration allow energy in glucose to be released slowly; therefore, ATP is produced gradually. In contrast, the rapid breakdown of glucose would result in the loss of most energy as unusable heat.

ATP can have three functions: (1) chemical work, where ATP supplies energy to synthesize molecules for the cell, (2) transport work, where ATP supplies energy to pump substances across the plasma membrane, and (3) mechanical work, where ATP supplies energy to perform muscle contraction, propel cilia, etc.

Oxidation-reduction half-reactions, electron carriers and flavoproteins

Oxidation-reduction reactions (redox reactions) involve the transfer of electrons. *Oxidation* is the loss of electrons, and *reduction* is the gain of electrons. Use the acronym "OIL RIG" (Oxidation Is Loss, Reduction Is Gain).

Although the focus of oxidation-reduction reactions is the transfer of electrons, other factors identify whether a reaction involves oxidation or reduction: oxidation frequently involves gaining oxygen or losing hydrogen, while reduction frequently involves losing oxygen or gaining hydrogen.

A half-reaction is either the oxidation or reduction component of a redox reaction. An oxidation reaction and its corresponding reduction reaction coincide because one molecule accepts the electrons given up by another molecule. In a half-reaction, the change in oxidation states in an individual substance is considered.

An example of an oxidation-reduction reaction is the reaction between fluorine and hydrogen, where fluorine is reduced (gaining electrons) and hydrogen is oxidized (losing electrons).

$$H_2 + F_2 \rightarrow 2\ HF$$

This redox reaction can be written as two half-reactions. The oxidation reaction is:

$$H_2 \rightarrow 2\ H^+ + 2\ e^-$$

The reduction reaction is:

$$F_2 + 2\ e^- \rightarrow 2\ F^-$$

In biological systems, electron carriers are usually required to transport electrons from one molecule to another. The *electron donor* is the molecule that electron carriers accept electrons from, and the *electron acceptor* is the molecule that electron carriers give electrons to.

There are two major electron carriers in cells. One is nicotinamide adenine dinucleotide (NAD^+). When NAD^+ (the oxidized form) accepts a hydrogen ion (H^+) and two electrons ($2\ e^-$), it becomes reduced to $NADH + H^+$. The other important electron carrier is flavin adenine dinucleotide (FAD), which is a flavoprotein (it contains a nucleic acid derivative of riboflavin). FAD (the oxidized form) can accept two hydrogen ions ($2\ H^+$) and two electrons ($2\ e^-$) so that it is reduced to $FADH_2$. Both of these electron carriers are soluble in water.

Notes for active learning

Glycolysis, Gluconeogenesis and the Pentose Phosphate Pathway

Glycolysis as the first step of cellular respiration

In *cellular respiration*, cells release the energy in chemical bonds of food molecules and transfer this energy to ATP molecules, which allows for the efficient use of an organism's energy. The ATP generated during cellular respiration is used for life's essential processes.

Cellular respiration is *aerobic* (with oxygen) or *anaerobic* (without oxygen).

Glycolysis (*glyco*l for sugar, *lysis* for breaking) is the first step in cellular respiration and the same in aerobic and anaerobic cells because glycolysis does not require oxygen. Glycolysis occurs in the cytosol of the cell and catabolizes a six-carbon glucose molecule into two three-carbon pyruvate molecules.

During glycolysis, energy is transferred through phosphate groups undergoing hydrolysis reactions (breaking) and condensation (joining).

The overall reaction of glycolysis is:

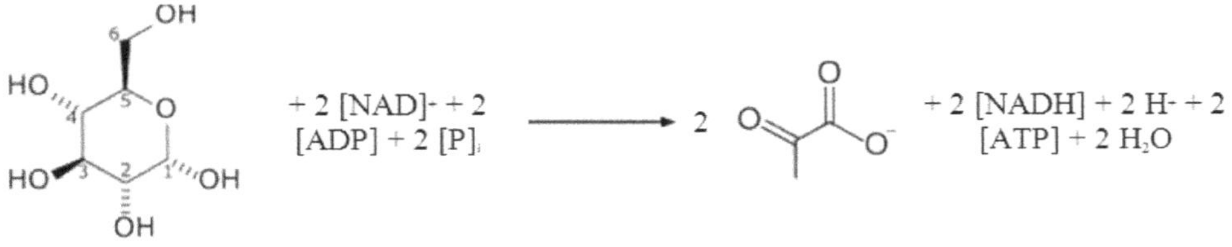

$$D\text{-}Glucose + 2\ [NAD]^- + 2\ [ADP] + 2\ [P]_i \longrightarrow 2\ Pyruvate + 2\ [NADH] + 2\ H^- + 2\ [ATP] + 2\ H_2O$$

D-Glucose   *Pyruvate*

Glycolysis yields two ATP molecules and two NADH molecules per molecule of glucose.

While it consists of ten steps, the summary of the steps is:

1. The glucose molecule undergoes several enzymatically regulated reactions and becomes a double-phosphorylated fructose molecule.

2. The 6-carbon fructose molecule is cleaved into two glyceraldehyde 3-phosphates (3-carbon molecules with a phosphate group), abbreviated GA3P (or PGAL). Transformation requires 2 ATP.

3. Hydrogen and water are removed from two PGAL, leaving two pyruvate molecules. This creates 2 NADH and 4 ATP (net ATP production is 2 ATP).

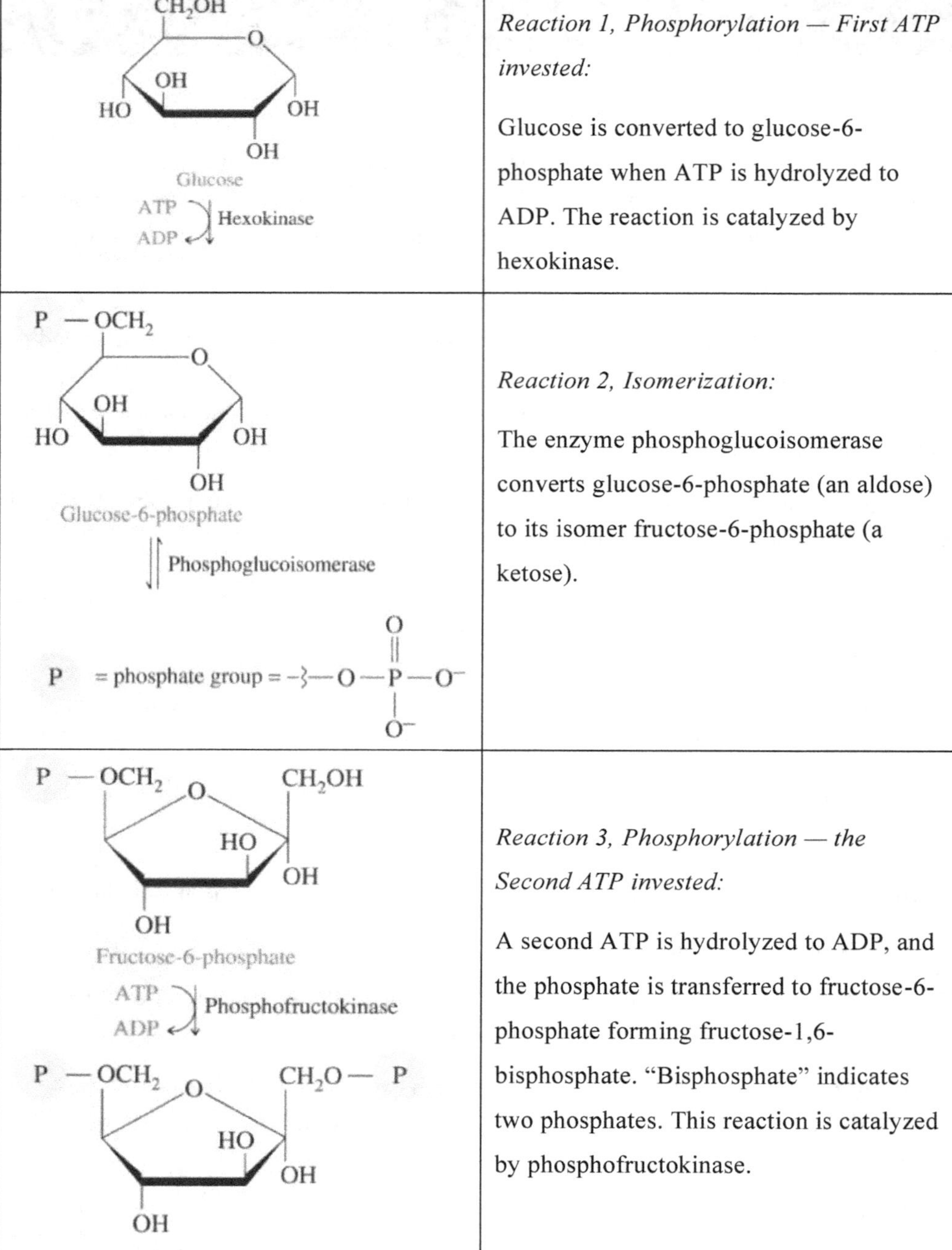

	Reaction 1, Phosphorylation — First ATP invested:
	Glucose is converted to glucose-6-phosphate when ATP is hydrolyzed to ADP. The reaction is catalyzed by hexokinase.
	Reaction 2, Isomerization:
	The enzyme phosphoglucoisomerase converts glucose-6-phosphate (an aldose) to its isomer fructose-6-phosphate (a ketose).
	Reaction 3, Phosphorylation — the Second ATP invested:
	A second ATP is hydrolyzed to ADP, and the phosphate is transferred to fructose-6-phosphate forming fructose-1,6-bisphosphate. "Bisphosphate" indicates two phosphates. This reaction is catalyzed by phosphofructokinase.

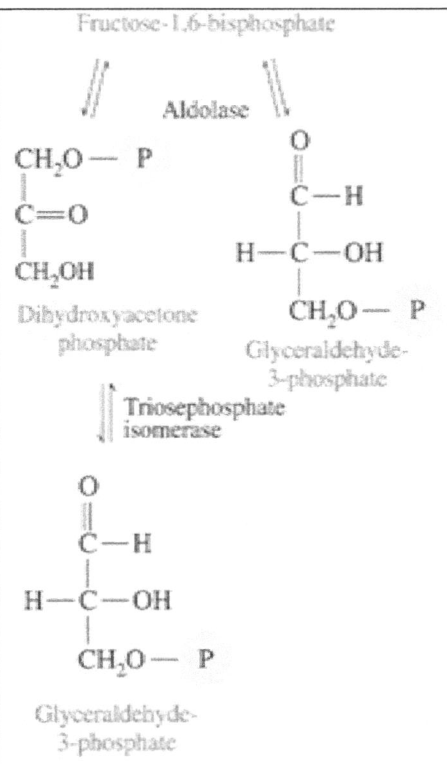

Reaction 4, Cleavage — Two trioses are formed:

Fructose-1,6-bisphosphate is cleaved into two triose phosphates (dihydroxyacetone phosphate and glyceraldehyde-3-phosphate), catalyzed by aldolase.

Reaction 5, Isomerization of a triose:

Triosephosphate isomerase converts one of the triose products (hydroxyacetone phosphate) to the other (glyceraldehyde-3-phosphate). Now, all 6 carbon atoms from glucose are in two identical 3-carbon triose phosphates.

Reaction 6, First energy production yields NADH:

The aldehyde group of glyceraldehyde-3-phosphate is oxidized and phosphorylated by the enzyme glyceraldehyde-3-phosphate dehydrogenase. The coenzyme NAD^+ is reduced to the high-energy compound NADH (and H^+) in the process.

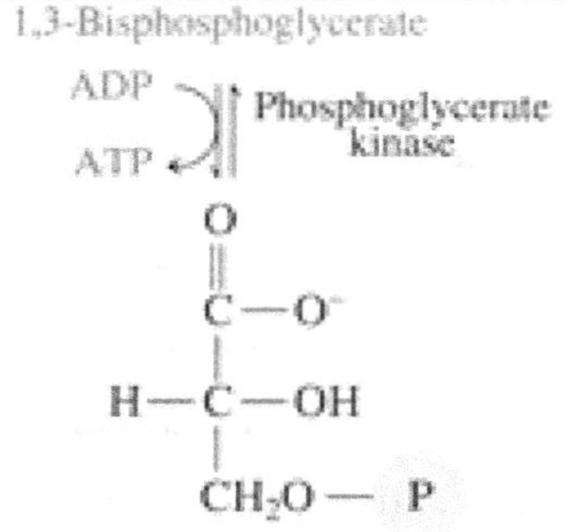 1,3-Bisphosphoglycerate ADP, ATP — Phosphoglycerate kinase $$\begin{array}{c} O \\ \| \\ C-O^- \\ \| \\ H-C-OH \\ \| \\ CH_2O-\text{P} \end{array}$$	*Reaction 7, Next energy production yields ATP:* The energy-rich 1,3-bisphosphoglycerate drives the formation of ATP when phosphoglycerate kinase transfers one phosphate from 1,3-bisphosphoglycerate to ADP.
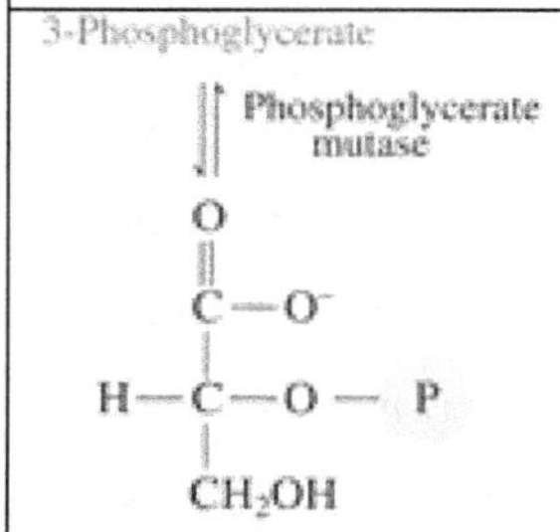 3-Phosphoglycerate Phosphoglycerate mutase $$\begin{array}{c} O \\ \| \\ C-O^- \\ \| \\ H-C-O-\text{P} \\ \| \\ CH_2OH \end{array}$$	*Reaction 8, Formation of 2-phosphoglycerate:* A phosphoglycerate mutase transfers the phosphate group from carbon 3 to carbon 2 to yield 2-phosphoglycerate.
2-Phosphoglycerate H_2O — Enolase $$\begin{array}{c} O \\ \| \\ C-O^- \\ \| \\ C-O-\text{P} \\ \| \| \\ CH_2 \end{array}$$	*Reaction 9, Removal of water makes a high-energy enol:* Enolase catalyzes the removal of water to yield phosphoenolpyruvate, a high-energy compound that transfers its phosphate in the next step.
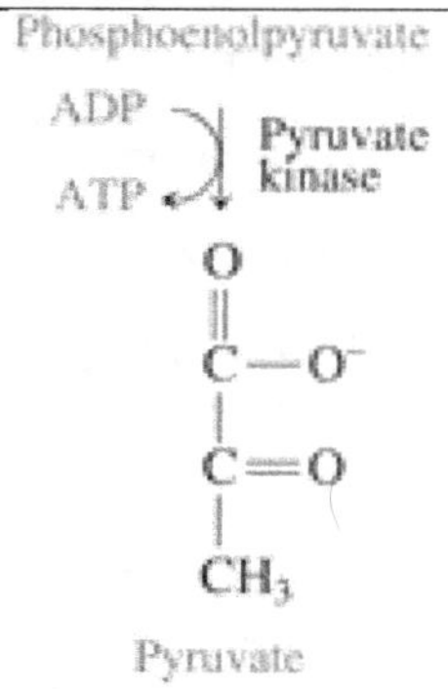 Phosphoenolpyruvate ADP, ATP — Pyruvate kinase $$\begin{array}{c} O \\ \| \\ C-O^- \\ \| \\ C=O \\ \| \\ CH_3 \end{array}$$ Pyruvate	*Reaction 10, Third energy production yields a second ATP:* ATP is generated in this final reaction when the phosphate from phosphoenolpyruvate is transferred. The pyruvate kinase catalyzes the reaction.

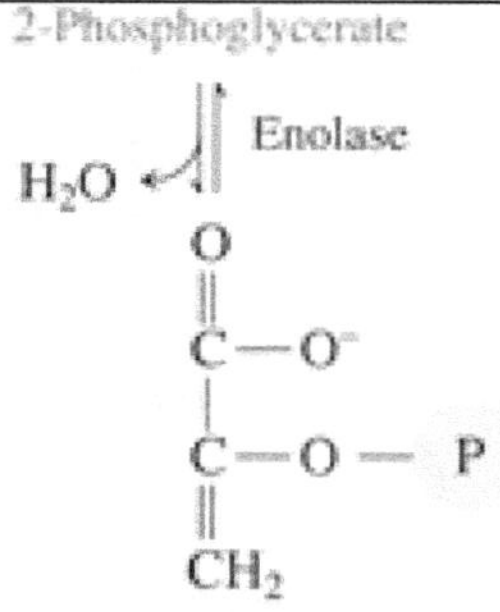

Glucose is not always immediately available, as it is stored in skeletal muscles and the liver as the polysaccharide glycogen. Hormones control whether glucose enters the anabolic pathway to form glycogen or the catabolic pathway to undergo glycolysis to form pyruvate. If the catabolic pathway signals are initiated, cellular respiration begins (glycolysis → Krebs → electron transport chain).

With the help of glycogen phosphorylase, glycogen can be converted into glucose 6-phosphate, and enter the second step in the ten-step pathway of glycolysis.

Fructose can undergo glycolysis. In the liver, fructose catabolism is unregulated and can produce excess products that become stored in the body as fat. In the muscles, fructose is converted to fructose-6-phosphate, entering glycolysis at step 3. In the liver, it is converted to the trioses used in step 5 of glycolysis.

Starch (amylose and amylopectin) begins to be digested in the mouth by enzymes in the saliva. The enzyme α-amylase hydrolyzes some of the α-glycosidic bonds in the starch molecules, producing glucose, disaccharide maltose, and oligosaccharides.

Only monosaccharides are small enough to be transported from the gastrointestinal system into the bloodstream. To complete the digestion of starch, enzymes in the small intestine hydrolyze starch and disaccharides into monosaccharides.

The most important regulatory step in glycolysis is step 3. The enzyme phosphofructokinase, which catalyzes the phosphorylation of fructose-6-phosphate to fructose-1,6-bisphosphate, is tightly regulated by the cells because this step is irreversible and commits the pathway to glycolysis.

ATP acts as an inhibitor of phosphofructokinase, so if cells have sufficient ATP levels, glycolysis slows down. If there is not much ATP, then glycolysis continues. The step after glycolysis depends on the type of respiration (i.e., aerobic or anaerobic).

Fermentation produces lactic acid or ethanol

Glycolysis produces two pyruvate molecules, two NADH molecules, and two ATP molecules. At this point, it is possible to begin the aerobic part of cellular respiration (Krebs Cycle) or continue with anaerobic respiration.

Fermentation is the next step in anaerobic (i.e., without oxygen) cellular respiration, which occurs in the cytoplasm. It occurs in yeast and bacteria but occurs in oxygen-starved muscle cells of vertebrates.

The goal of fermentation reactions is to oxidize NADH produced in glycolysis into NAD^+ by reducing the pyruvate. The NAD^+ molecules are used in another round of glycolysis to produce two more ATP and two more NADH.

Fermentation is a slow and inefficient method of making ATP since only two ATP are produced in each cycle (compared to ~36 ATP produced in each aerobic respiration cycle).

Lactic acid fermentation produces two molecules of lactate (or lactic acid as the acidic form of the molecule) and occurs in bacteria and some fungi. It is used to produce some foods (e.g., yogurt).

Lactic acid fermentation occurs in humans and other mammals' muscle cells during demanding physical activities (e.g., sprinting) when energy demands are high; lactic acid fermentation provides a quick burst of energy via ATP synthesis needed for the muscular activity.

L Lactic acid is toxic to mammals; this is the "burn" felt when undergoing strenuous activity. When blood cannot remove lactate from muscles, this decreases the pH, causing muscle fatigue.

Oxygen debt is the oxygen that the body needed but is not delivered to the cell. Oxygen is needed to restore ATP levels and rid the body of lactate ("repaying" the oxygen debt), which is why a person might breathe harder after exercise. Recovery occurs after lactate is sent to the liver, converted into pyruvate; some pyruvate is expired or converted into glucose.

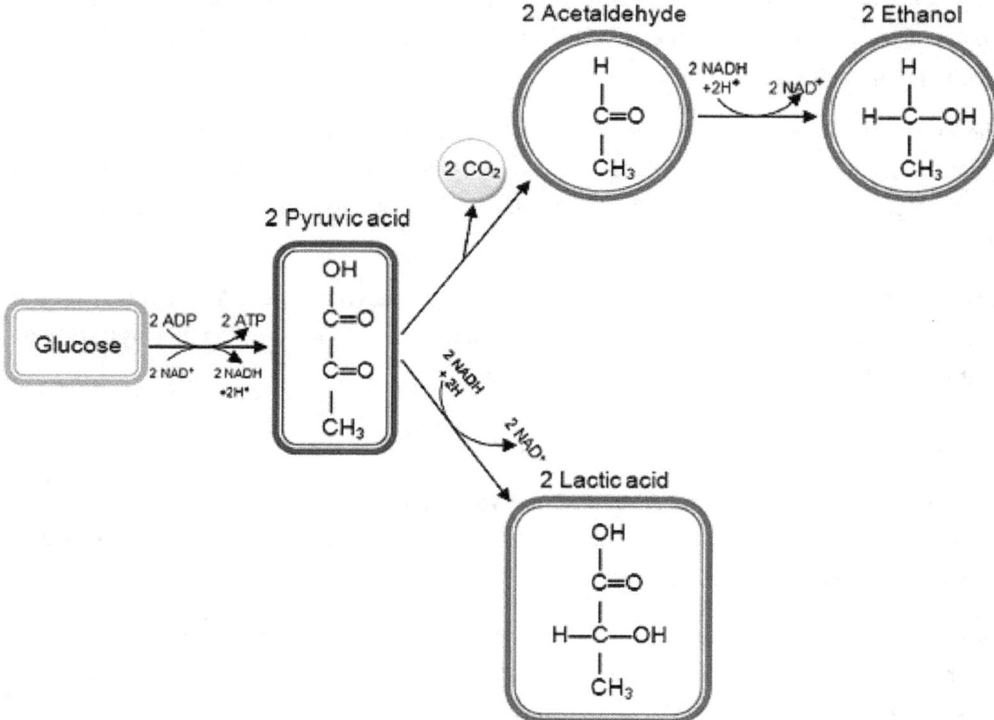

Alcoholic fermentation for ethanol production and lactic acid fermentation pathways to regenerate NAD⁺

During lactic acid fermentation, the middle carbonyl (C=O) in pyruvate is reduced (i.e., hydrogen is added) to an OH group, and lactate is formed, as catalyzed by lactate dehydrogenase.

The hydrogen (and energy) required for this reaction is supplied by NADH, producing NAD^+. NADH passes its electrons to pyruvate, and NAD^+ then returns to the glycolysis pathway to receive more electrons. In lactic acid fermentation, pyruvate is the final electron acceptor.

Alcoholic fermentation is the production of two ethanol molecules and two carbon dioxide molecules. Yeasts and some types of bacteria perform alcoholic fermentation. Some yeasts perform aerobic cellular respiration when oxygen is available, and only utilize alcoholic fermentation in anaerobic environments.

However, many yeasts prefer fermentation, even if oxygen is available. These yeasts are utilized in the production of bread and alcoholic beverages. *Saccharomyces cerevisiae, the* yeast used in baking, consumes sugars in the dough of bread (converting glucose to pyruvate during glycolysis) and then reduces the pyruvate, creating carbon dioxide and ethanol as waste products.

The carbon dioxide causes the dough to rise, and the ethanol evaporates when the bread is baked. *S. cerevisiae* is used in the production of beer. This yeast consumes the grain starches during glycolysis, and the carbon dioxide that is produced along with the ethanol during alcoholic fermentation is carbonation in the beer. Just as lactic acid is toxic to mammals, ethanol is toxic to the microorganisms that produce it.

Unlike lactic acid fermentation, where the pyruvate directly accepts electrons from NADH, there is an intermediate compound in alcoholic fermentation. The two pyruvate molecules are first converted into two molecules of acetaldehyde, catalyzed by pyruvate decarboxylase. The byproduct is carbon dioxide.

The acetaldehyde is the final electron acceptor (from NADH). Then, the two acetaldehyde molecules are converted into two molecules of ethanol, catalyzed by the enzyme alcohol dehydrogenase.

$$\underset{\textit{Pyruvate}}{CH_3-\overset{\overset{O}{\|}}{C}-\overset{\overset{O}{\|}}{C}-O^-} + NADH + 2H^+ \longrightarrow \underset{\textit{Ethanol}}{CH_3-\overset{\overset{HO}{|}}{\underset{\underset{H}{|}}{C}}-H} + CO_2 + NAD^+$$

Gluconeogenesis

In addition to catabolizing glucose, many organisms produce it from non-carbohydrate substances, as *gluconeogenesis*. Gluconeogenesis occurs in the mitochondria and cytoplasm of many organisms, including animals, plants, fungi, and bacteria. For vertebrates, gluconeogenesis occurs primarily in the liver and, to a limited extent, in the kidneys. This process maintains the glucose concentration in the blood.

A variety of non-carbohydrate carbon substrates, including pyruvate, glycerol, lactate, and certain amino acids, are used as the starting molecule in gluconeogenesis. If the starting molecule is not pyruvate, the first step in gluconeogenesis is to convert the precursor (e.g., lactate) to pyruvate. An amino acid precursor enters the gluconeogenesis metabolic pathway at oxaloacetate or later in the pathway (for glycerol).

The steps of gluconeogenesis are displayed below, along with enzymes that catalyze each step. The entire metabolic pathway of gluconeogenesis uses two ATP, two GTPs and one NADH. As seen from the similarity of the two pathways, gluconeogenesis can be thought of as "reverse glycolysis." However, three of the enzymes used in the two pathways are different, so the energy cost of gluconeogenesis is not too great to be energetically favorable for the organism. Instead of using hexokinase, phosphofructokinase and pyruvate kinase (used in glycolysis), gluconeogenesis uses glucose-6-phosphatase, fructose-1,6-bisphosphatase and PEP carboxykinase/pyruvate carboxylase. This replaces three highly endergonic reactions in glycolysis with reactions that are exergonic and therefore favorable.

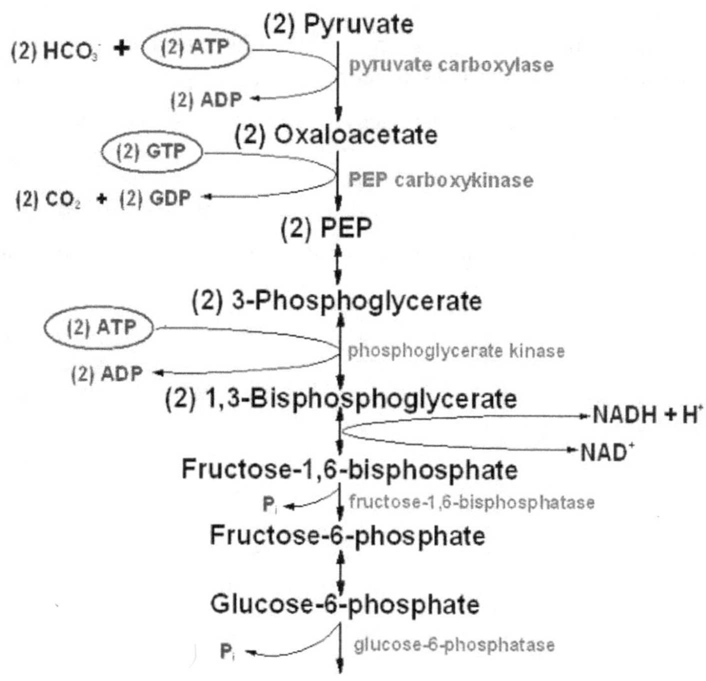

Glucose is produced via gluconeogenesis

In addition to building up glucose from carbon substrates (i.e., gluconeogenesis), organisms derive glucose by breaking down stored energy sources, such as the polysaccharides of starch and glycogen, with hormones controlling these processes.

The hormone glucagon promotes glycogen degradation and inhibits glycolysis in the liver, which causes the glycolytic intermediates to be used in gluconeogenesis. Glucagon does this by inhibiting the phosphofructokinase (PFK) enzyme, since PFK is integral to glycolysis, to increase the concentration of glucose in the blood. Conversely, insulin inhibits gluconeogenesis by activating the PFK enzyme, so that the concentration of glucose in the blood is reduced.

Pentose phosphate pathway

The *pentose phosphate pathway* is a metabolic pathway that produces nicotinamide adenine dinucleotide phosphate (NADPH) and five-carbon sugars. In most organisms, it takes place in the cytosol, the exception being plants where it occurs in plastids. This pathway can be thought of as "parallel" to glycolysis, but rather than being catabolic, the pentose phosphate pathway is anabolic, since it synthesizes five-carbon pentose sugars.

The first phase of the pentose phosphate pathway is the oxidative phase, where two $NADP^+$ molecules are reduced to two NADPH molecules, and glucose-6-phosphate is oxidized to ribulose-5-phosphate. The production of NADPH in the pentose phosphate pathway is vital because it is the primary source of NADPH in non-photosynthetic organisms.

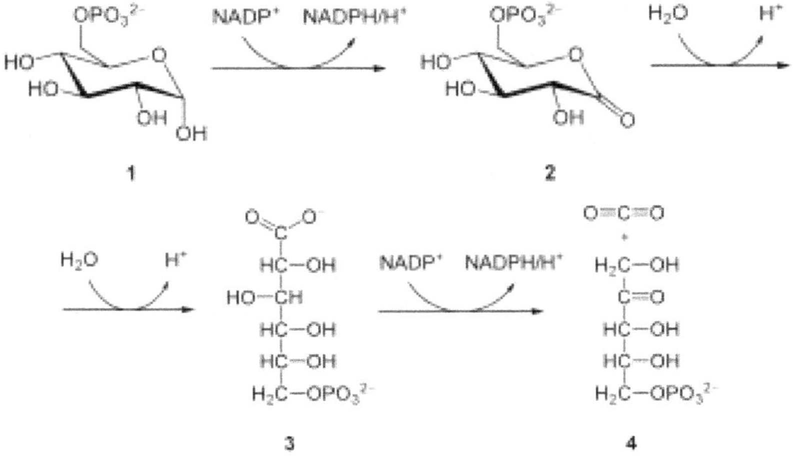

The oxidative phase of the pentose phosphate pathway

1: glucose-6-phosphate

2: 6-phosphogluconolactone

3: 6-phosphogluconate

4: ribulose 5-phosphate

The second phase of the pentose phosphate pathway is the non-oxidative phase, where other types of pentoses are synthesized from ribulose-5-phosphate. One of these pentoses is ribose-5-phosphate, which is used to synthesize nucleic acids. Another type of pentose is erythrose-4-phosphate, which is used to generate aromatic amino acids. The pentose phosphate pathway is important, as it enables the biosynthesis of the biomolecules as the building blocks of life.

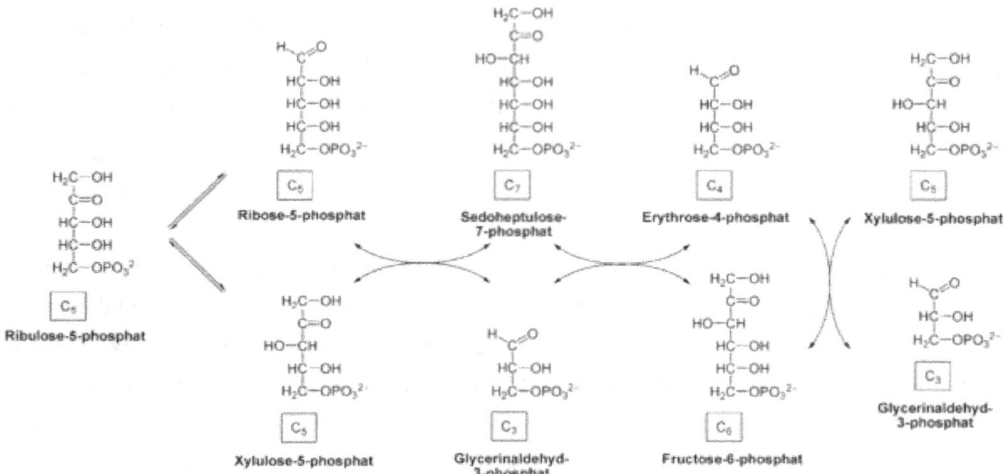

The second phase of pentose phosphate pathway with ribose-5-phosphate (nucleic acids) and erythrose-4-phosphate. Note the spelling of phosphate should include an -e ending

Net molecular and energetic results of respiration processes

From fermentation, the two ATP produced per glucose molecule is equivalent to 14.6 kcal. The ATP is produced via substrate-level phosphorylation as the direct enzymatic transfer of a phosphate to ADP, no extraneous carriers needed. Complete glucose breakdown to CO_2 and H_2O during aerobic cellular respiration represents a possible yield of 686 kcal of energy.

Therefore, efficiency for fermentation is 14.6 / 686, or about 2.1%– much less efficient than aerobic respiration. Thus, the presence of an oxygen-rich atmosphere, which facilitated the evolution of aerobic respiration, is crucial in the complexity and diversification of life.

Cellular Respiration Reactions

Acetyl-CoA production

Aerobic cellular respiration includes the anaerobic process of glycolysis and the aerobic processes in the mitochondria of eukaryotes.

Three processes within the mitochondria are pyruvate decarboxylation, the Krebs cycle (the citric acid cycle or TCA), and oxidative phosphorylation via the electron transport chain (ETC).

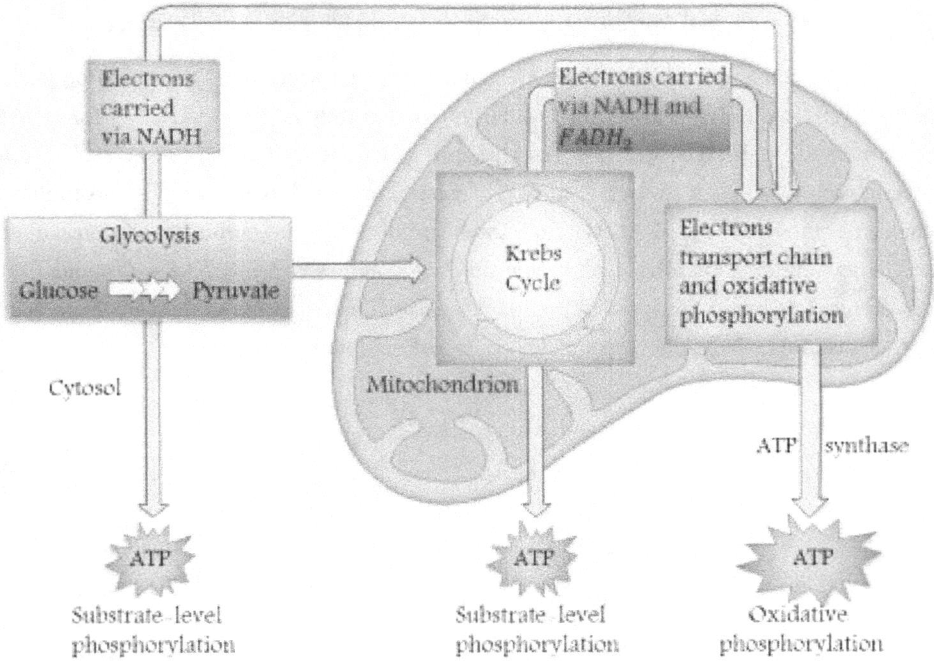

Cellular respiration for glycolysis, Krebs cycle, and electron transport chain

Pyruvate decarboxylation follows glycolysis in aerobic respiration.

Pyruvate decarboxylation is a link reaction as the intermediate step after glycolysis and before the Krebs cycle, thereby linking the two metabolic pathways.

In eukaryotes, pyruvate decarboxylation occurs in the mitochondrial matrix (cytosol of mitochondria). In prokaryotes, it occurs in the cytoplasm and at the plasma membrane. It converts pyruvate into acetyl coenzyme A (acetyl-CoA) and is catalyzed by the pyruvate dehydrogenase complex (PDC), a complex of three enzymes. From two pyruvate molecules (one original molecule of glucose), two acetyl-CoA are created, and two NAD^+ are reduced to NADH. Additionally, two molecules of CO_2 are released.

Coenzyme A (CoA) is an important energy exchanger that contains adenosine, three phosphates, and a pantothenic acid-derived (vitamin B_5) portion. The two forms are acetyl-CoA (high energy) and CoA (low energy). Energy is released from acetyl-CoA when the C–S bond in the thioester group is hydrolyzed, producing an acetyl group and CoA.

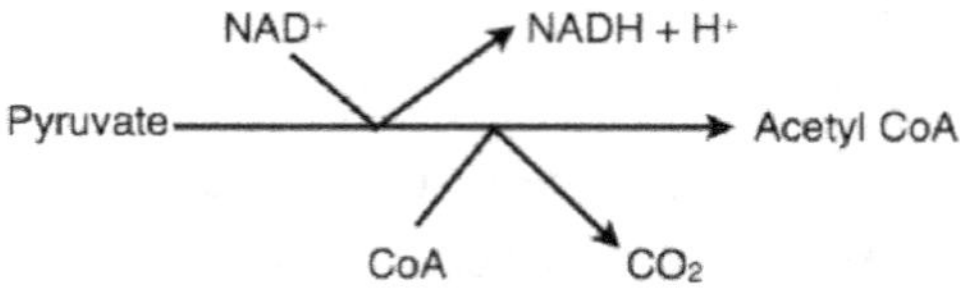

Coenzyme A exists in a low energy form as CoA or high energy form as acetyl-CoA

In addition to glucose breakdown, there are other ways that acetyl-CoA can be produced for use in the Krebs cycle. Fat breaks down into glycerol and fatty acids. Glycerol is converted to PGAL, a metabolite in glycolysis. Beta oxidation is when the fatty acids can be broken down to generate acetyl-CoA. An 18-carbon fatty acid can be converted to nine acetyl-CoA molecules that each enter the Krebs cycle. Acetyl-CoA can be produced by degrading the carbon skeletons of ketogenic amino acids. However, regardless of whether acetyl-CoA is produced from carbohydrates, fats or proteins, the subsequent step for this molecule is the same: the acetyl-CoA proceeds to the Krebs cycle as the next phase of aerobic respiration.

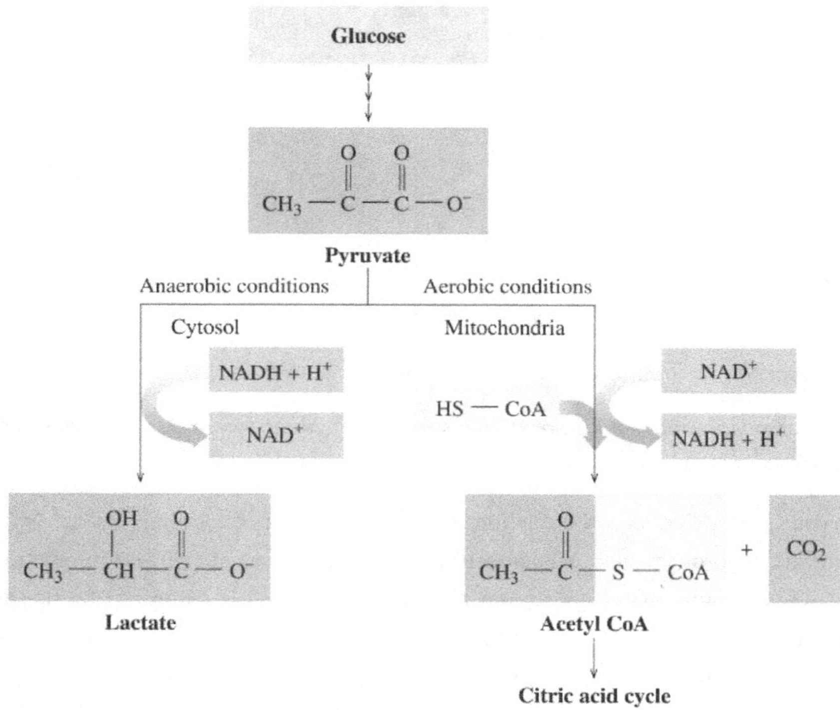

Pyruvate is the product of glycolysis that proceeds into the citric acid (Krebs) cycle or
fermentation when oxygen is absent

Krebs cycle substrates and products

The *Krebs cycle* occurs in the fluid matrix of the mitochondria's cristae compartments. The cycle is named after Sir Hans Krebs, who received the 1953 Nobel Prize for identifying these reactions.

The Krebs (or *citric acid*) cycle is the tricarboxylic acid cycle (TCA) cycle because of intermediate acids.

The Krebs cycle removes energy, carbon dioxide, and hydrogen from acetyl-CoA via enzyme-mediated reactions of organic acids. It begins by joining acetyl-CoA (2 carbons) with oxaloacetate (4 carbons), producing citric acid (6 carbons).

Citric acid undergoes several oxidations, decarboxylation, dehydrogenation, and hydration, yielding two CO_2, one GTP, three NADH, and one $FADH_2$ for each turn.

One glucose yields four CO_2, two GTP, six NADH, and two $FADH_2$.

GTP readily converts to ATP in the cell.

The cycle regenerates oxaloacetate to begin again. Glucose splits into two pyruvates during glycolysis; one glucose undergoes two turns of the Krebs cycle.

The oxidations release energy stored by the nucleotide carriers (NADH and $FADH_2$) when they accept the hydrogen electrons.

Cytochromes use this stored energy in NADH and $FADH_2$ in the electron transport chain to produce ATP by oxidative phosphorylation.

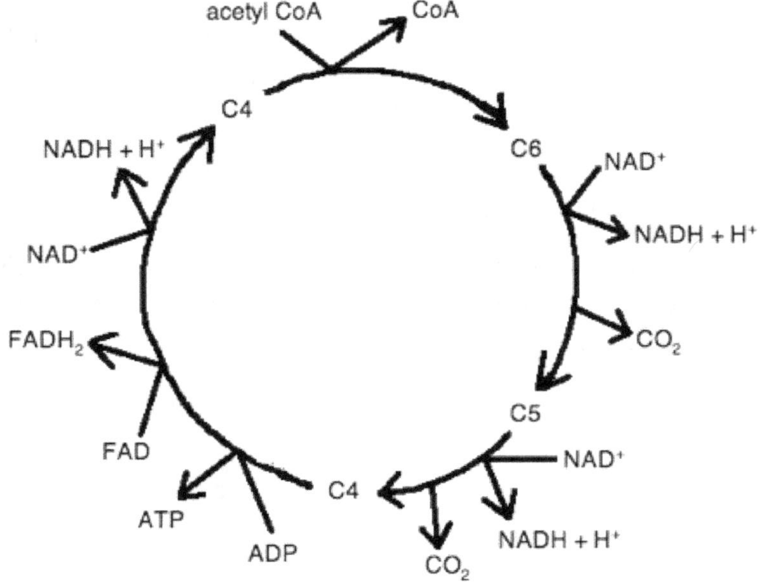

Krebs cycle showing the reactants and products. GTP is equivalent to ATP

The details of the Krebs cycle are as follows:

Reaction 1, Formation of Citrate: The acetyl group from acetyl-CoA (two carbons) combines with oxaloacetate (four carbons), forming citrate (six carbons) and CoA.

Reaction 2, Isomerization to Isocitrate: The OH and one of the H atoms are exchanged in citrate to form isocitrate. This rearrangement is necessary because isocitrate is oxidized in the next reaction.

Reaction 3, First Oxidative Decarboxylation (Release of CO_2): An alcohol undergoes oxidation (electron and two hydrogens are removed) to the ketone α-ketoglutarate and NAD^+ is reduced to NADH, accepting the proton and electrons removed during the oxidation. The six-carbon isocitrate is decarboxylated (release of CO_2) to the five-carbon α-ketoglutarate.

Reaction 4, Second Oxidative Decarboxylation: The thiol group of CoA is oxidized (loses an electron), and another NAD^+ is reduced (gains electron) to NADH. α-ketoglutarate (five carbons) is decarboxylated into a succinyl group (four carbons). The CoA is bonded to the succinyl group, thus producing succinyl CoA.

Reaction 5, Hydrolysis of Succinyl CoA: Succinyl CoA undergoes hydrolysis to yield succinate and CoA. The resulting energy produces the high-energy nucleotide GTP (guanosine triphosphate) from GDP and P_i. The GTP is converted to ATP in the cell.

Reaction 6, Dehydrogenation of Succinate: One hydrogen is eliminated from each of the two central carbons of succinate, forming a *trans* C=C bond, thus producing fumarate. The coenzyme FAD is reduced to $FADH_2$.

Reaction 7, Hydration of Fumarate: Water adds to the *trans* double bond of fumarate as H and OH to form malate.

Reaction 8, Oxidation of Malate: As in reaction 3, the secondary alcohol of malate is oxidized to a ketone, forming oxaloacetate and providing protons and electrons for reducing the coenzyme NAD^+ to NADH.

Net molecular and energetic results of respiration processes

The two pyruvate molecules produced during glycolysis are converted into two acetyl-CoA molecules in the link reaction. The acetyl-CoA then enters the Krebs cycle, which turns twice because two acetyl-CoA molecules enter the cycle per original glucose molecule.

The final products of the Krebs cycle are oxaloacetic acid (to further drive the cycle), 2 ATP (converted from GTP), 6 NADH, 2 FADH$_2$, and 4 CO$_2$. The high-energy molecules NADH and FADH$_2$ are used in the final step of aerobic respiration, oxidative phosphorylation.

The equation for one turn of the Krebs cycle:

$$\text{Acetyl-CoA} + 3\ \text{NAD}^+ + \text{FAD} + \text{GDP} + \text{P}_i + 2\ \text{H}_2\text{O} \rightarrow 2\ \text{CO}_2 +$$
$$3\ \text{NADH} + 2\ \text{H}^+ + \text{FADH}_2 + \text{CoA} + \text{GTP}$$

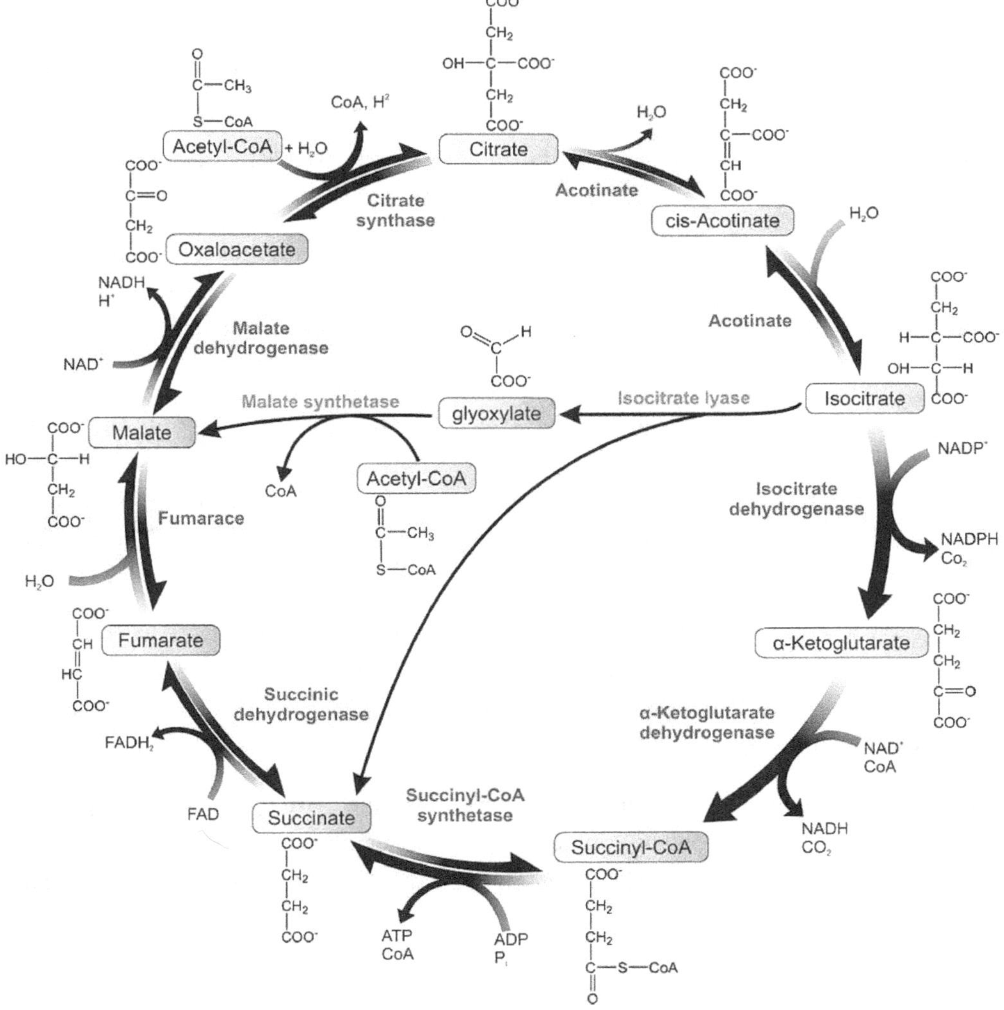

Krebs cycle with structures, enzymes and produces

The net production in cellular respiration from glycolysis through the Krebs cycle is 8 NADH, 2 $FADH_2$, 2 ATP, and 6 CO_2.

To summarize, the Krebs cycle degrades two-carbon acetyl groups from acetyl CoA into CO_2, and the carbon dioxide is released in two reactions (as a waste product exhaled in animals).

GTP is produced in one of the reactions and is converted into ATP.

Hydrogen is removed in four reactions. NAD^+ accepts two electrons and the proton in three reactions, creating three NADH.

FAD accepts two electrons and the proton in one reaction, creating one $FADH_2$.

Krebs cycle regulation

The Krebs cycle must be carefully regulated to generate the proper amount of ATP. Although oxygen is not directly used in the Krebs cycle, the cycle can only occur under aerobic conditions because FAD and NAD^+ can be regenerated when oxygen is present. If oxygen is drastically reduced, respiration may drop to the point where it may lead to death.

Low oxygen concentrations divert aerobic cellular respiration from the Krebs cycle to anaerobic fermentation.

The Krebs cycle is mainly regulated by substrate availability, product inhibition, and competitive feedback inhibition. ADP (a substrate) is converted to ATP during the cycle, and a decreased amount of ADP reduces the cycle rate. This leads to NADH accumulation, decreasing the amount of NAD^+ available for the cycle.

NADH can allosterically inhibit the enzymes in the Krebs cycle, including the pyruvate dehydrogenase complex (PDC), which catalyzes the link reaction (i.e., pyruvate decarboxylation to produce acetyl-CoA). Acetyl-CoA inhibits the PDC as an example of end-product inhibition (or feedback inhibition) because acetyl-CoA is the product of the reaction catalyzed by the PDC.

Acetyl-CoA can enter the Krebs cycle from sources other than glycolysis, such as the breakdown of fatty acids. Calcium is an activator of the PDC and activates other dehydrogenase enzymes that catalyze reactions in the Krebs cycle. AMP (adenosine monophosphate) activates the PDC.

Citrate, the first compound in the Krebs cycle (formed from acetyl-CoA joining oxaloacetate), is used for feedback inhibition. Citrate inhibits phosphofructokinase, an important enzyme used in glycolysis. When high quantities of citrate accumulate, the rate of the respiration pathway is reduced to prevent the overproduction of ATP.

Oxidative Phosphorylation

Oxidative phosphorylation and the electron transport chain

Oxidative phosphorylation is the next step in aerobic cell respiration, including the electron transport chain (ETC). This step produces the majority of ATP. This is done by the oxidation (loss of electrons) by high-energy intermediates (NADH and $FADH_2$), causing H^+ to be pumped into the intermembrane space (between inner and outer membranes) of the mitochondria.

The electron transport chain reactions occur at the *cristae*, which are the folds of the inner mitochondrial membrane that increase the surface area for the electron transport chain. During this process, a gradient across the mitochondrial membrane is created to drive ATP production.

Depending on cell conditions and if involving a prokaryotic or eukaryotic organism, about 32 to 34 molecules of ATP are produced by oxidative phosphorylation per glucose molecule.

The *electron transport chain* uses the NADH and $FADH_2$ from the Krebs cycle for a series of protein complexes that extract energy and pump protons across the inner mitochondrial membrane.

Energy is released as the hydrogen and electrons from the NAD^+ and FAD^+ carrier molecules flow into the system.

When the electrons reach the end of the chain, they are accepted by oxygen (the final electron acceptor), and water is released when oxygen combines with the electrons and protons.

Chemiosmosis is the mechanism of ATP generation that occurs when energy is stored in the form of a proton concentration gradient across a membrane. ATP is produced by oxidative phosphorylation during the action of the electron transport chain.

When H^+ molecules from the carrier molecules are transported from the matrix to the intermembrane space, a pH and electric charge gradient are created. The ATP synthase enzyme uses the potential energy on this gradient to create ATP when the protons flow through the ATPase channel in the membrane back into the matrix.

Note: A common question concerns pH changes from these processes. Remember that an *increase* in H^+ concentration means a *decrease* in pH.

Within the inner membrane of the mitochondria, there are four enzyme complexes (numbered I – IV). ATP synthase is sometimes referred to as complex V. The electron carriers' coenzyme Q (CoQ) and cytochrome c are not firmly attached to any complex and shuttle electrons between the complexes.

CoQ (known as ubiquinone) is a fat-soluble carrier dissolved in the membrane. It can be fully reduced, fully oxidized, or somewhere in between. This property enables it to perform in the electron transport chain. CoQ carries electrons from Complex I and Complex II to Complex III.

Like all coenzymes, CoQ has a vitamin-like structure. *Cytochrome c* is a water-soluble protein that transfers electrons between Complex III and Complex IV. There is a central iron atom in this molecule; it is surrounded by heme protein. Cytochrome c is highly conserved across species, and it is often used to study evolutionary relationships between organisms.

The details of the electron transport chain are as folows.

> *Complex I, NADH Dehydrogenase:* NADH enters the electron transport chain. During its oxidation, two electrons and two protons are transferred to the electron transporter CoQ, reducing its two ketone groups to alcohols. NAD^+ is regenerated and returns to a catabolic pathway, as in the Krebs cycle. The reaction at Complex I is $NADH + H^+ + Q \rightarrow NAD^+ + QH$.

> *Complex II, Succinate Dehydrogenase:* $FADH_2$ enters electron transport after the reduced nucleotide is produced in the conversion of succinate to fumarate in the Krebs cycle. Two electrons and two protons from $FADH_2$ are transferred to CoQ to yield QH_2. The reaction at Complex II is $FADH_2 + Q \rightarrow FAD + QH_2$.

> *Complex III, Coenzyme Q–Cytochrome c Reductase:* The reduced coenzyme Q (QH_2) molecules are reoxidized to ubiquinone (Q). The electrons pass through a series of electron acceptors until they arrive at cytochrome c, which moves the electrons from Complex III to Complex IV.

> *Complex IV, Cytochrome c Oxidase:* Single electrons are transferred from cytochrome c through another set of electron acceptors to combine with hydrogen ions and oxygen as the final electron accepts to form water. The reaction is $4 H^+ + 4 e^- + O_2 \rightarrow 2 H_2O$.

Three of the complexes (I, III and IV) span the inner membrane and pump protons out of the matrix and into the intermembrane space as electrons are shuttled through the complexes. The only complex that does not pump protons is Complex II.

The formation of the chemiosmotic (proton) gradient across the inner mitochondrial membrane provides the energy for ATP synthesis. Protons move back into the matrix through the transmembrane ATP synthase enzyme, and the resulting release of potential energy from the chemiosmotic gradient drives the synthesis of ATP by oxidative phosphorylation as protons and electrons join $\frac{1}{2}O_2$.

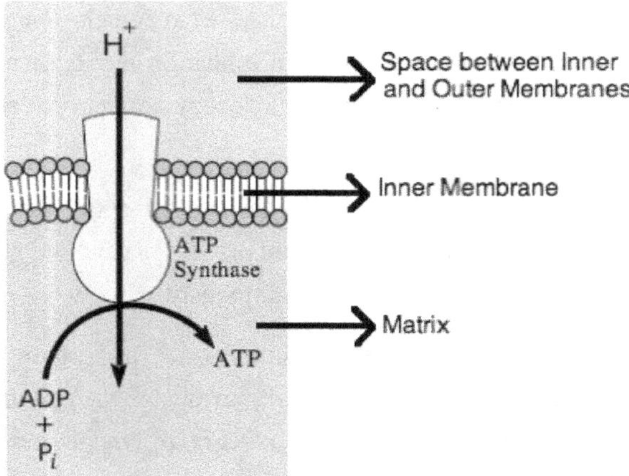

Oxidative phosphorylation with protons passing through the ATP synthase

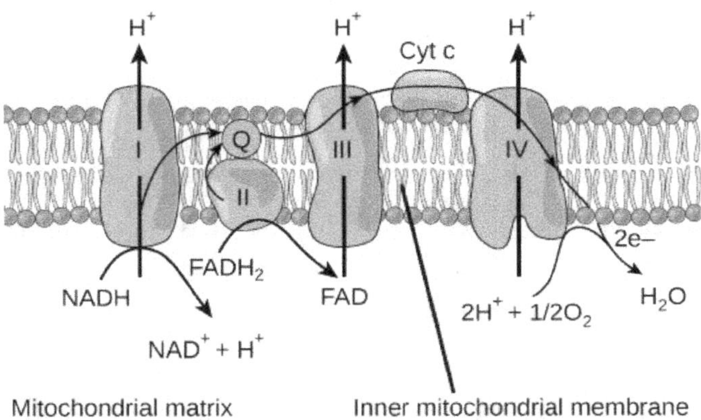

Electron transport chain showing the membrane-bound cytochromes as electrons move down their reduction potential towards O₂ as the final electron acceptor

Electron transfer in mitochondria: NADH, NADPH, flavoproteins, and cytochromes

The coenzymes nicotinamide adenine dinucleotide (NAD^+) and flavin adenine dinucleotide (FAD) are energy-transferring compounds that exist in different redox states. NAD^+ (low-energy oxidized form) accepts two electrons and proton to become NADH (high-energy reduced form).

FAD (low-energy oxidized form) accepts two electrons and two protons to become $FADH_2$ (high-energy reduced form). The active end of each coenzyme contains a vitamin component. Nicotinamide is derived from the niacin (B_3) vitamin, and riboflavin (B_2) vitamin is found in FAD. The FAD is, therefore, a type of flavoprotein.

Electrons received by NAD^+ and FAD can then be carried to cytochrome protein of the electron transport chain. NAD^+ and FAD are coenzymes of oxidation-reduction since they both accept and donate electrons. Only a small amount of these coenzymes are needed in cells because the exchange of electrons regenerates each molecule.

For example, once NADH delivers electrons to the electron transport chain, it becomes oxidized to NAD^+ and then can be reduced with electrons and protons. Recycling NAD^+, FAD, and ADP eliminate the need to synthesize them *de novo* continuously.

NAD^+ is converted into nicotinamide adenine dinucleotide phosphate ($NADP^+$). $NADP^+$ is similar in structure to NAD^+ but has an additional phosphate group, and it can be reduced to NADPH. In non-photosynthetic organisms, the production of $NADP^+$ generally occurs through the pentose phosphate pathway.

In plants, however, it is produced during the last step of the electron transport chain (across the thylakoid membrane) in the light reactions of photosynthesis. In general, NAD^+, $NADP^+$ and FAD are reduced during catabolic processes; their reduced forms (NADH, NADPH, and $FADH_2$) are oxidized during anabolic processes.

For each NADH formed within the mitochondrion (during the link reaction and the Krebs cycle), three ATP are produced. For each $FADH_2$ formed by the Krebs cycle, two ATP are produced. This is because $FADH_2$ delivers electrons after NADH (Complex II vs. Complex I).

Therefore, NADH yields more energy than $FADH_2$, because more H^+ is pumped across the membrane per NADH (3:2 ratio). However, an exception occurs for NADH formed outside the mitochondrion (by glycolysis) in the cytoplasm.

The NADH formed by glycolysis is the same molecule as the NADH formed in the mitochondria but yields two ATP (equivalent to ATP production by FADH during the Krebs cycle) instead of three ATP. One ATP is consumed to transport the NADH of glycolysis from the cytoplasm into the mitochondrion, resulting in a net gain of two ATP.

ATP synthase and chemiosmotic coupling; proton motive force

The movement of protons from the mitochondrial matrix into the intermembrane space creates a concentration gradient. The *proton motive force* is the energy used to pump these protons across the inner membrane, and it comes from the energy released by the electrons passing through the electron transport chain.

Remember that only three out of four enzyme complexes in the electron transport chain can pump protons into the intermembrane space: complexes I, III and IV. The protons create a concentration gradient (lowering the pH) as they move into the intermembrane space.

Chemiosmosis is ATP production tied to an electrochemical (H^+) gradient across a membrane. There is now a high concentration of protons in the intermembrane space and a low concentration of protons in the mitochondrial matrix.

The protons then move down the concentration gradient from the intermembrane space back into the matrix. However, the inner membrane is impervious to protons, and they can move back into the matrix via the ATP synthase, an enzyme embedded in the inner membrane. ATP synthase complexes span the membrane and are channel proteins that serve as enzymes for ATP synthesis.

As the protons are transported back into the matrix through the channels of ATP synthase, they release energy, which is then used by ATP synthase to convert ADP into ATP by the addition of an inorganic phosphate (P_i).

This process is oxidative phosphorylation, because the electrons come from previous oxidation reactions of cell respiration, and the ATP synthase uses $\frac{1}{2}O_2$ to accept the electrons to catalyze the phosphorylation of ADP into ATP. Once formed, ATP molecules diffuse out of the mitochondrial matrix through channel proteins and become available for use by the cell.

Tests confirmed the chemiosmotic nature of ATP synthesis with respiratory poisons (i.e., poisons that inhibit ATP synthesis), where these poisons caused the H^+ concentration gradient to increase.

British biochemist Peter Mitchell received the 1978 Nobel Prize for his chemiosmotic theory of ATP production, which occurs in the membranes of both mitochondria in animal cells and chloroplasts in plant cells.

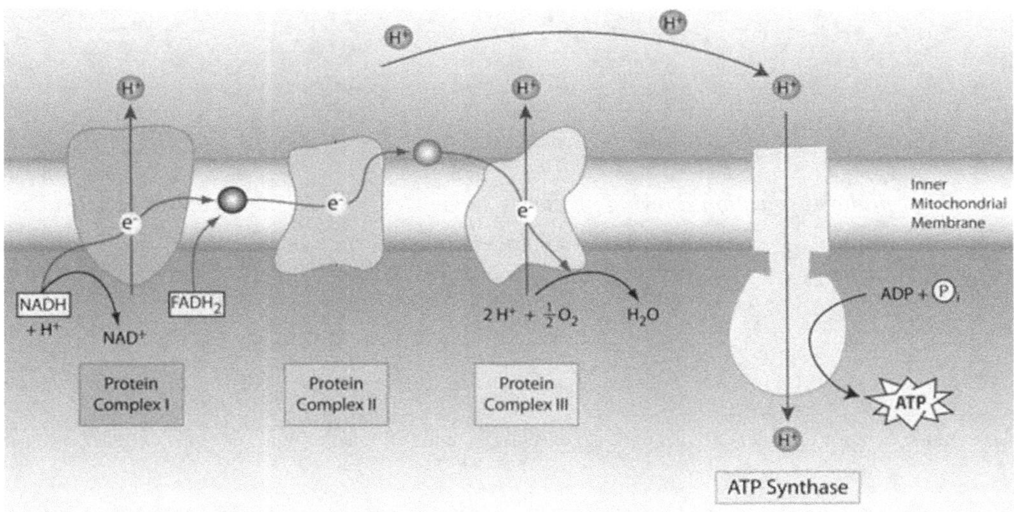

Inner membrane contains cytochromes that pass electrons. ATP synthase produces ATP by oxidative phosphorylation when protons pass down the concentration gradient into the matrix

The exergonic flow of electrons is generally coupled with the endergonic pumping of protons across the cristae membrane of the mitochondria. When electron flow and proton transport are uncoupled, the energy in the proton gradient is released as heat. This assists in maintaining body temperature through *thermogenesis*, which occurs primarily in warm-blooded animals.

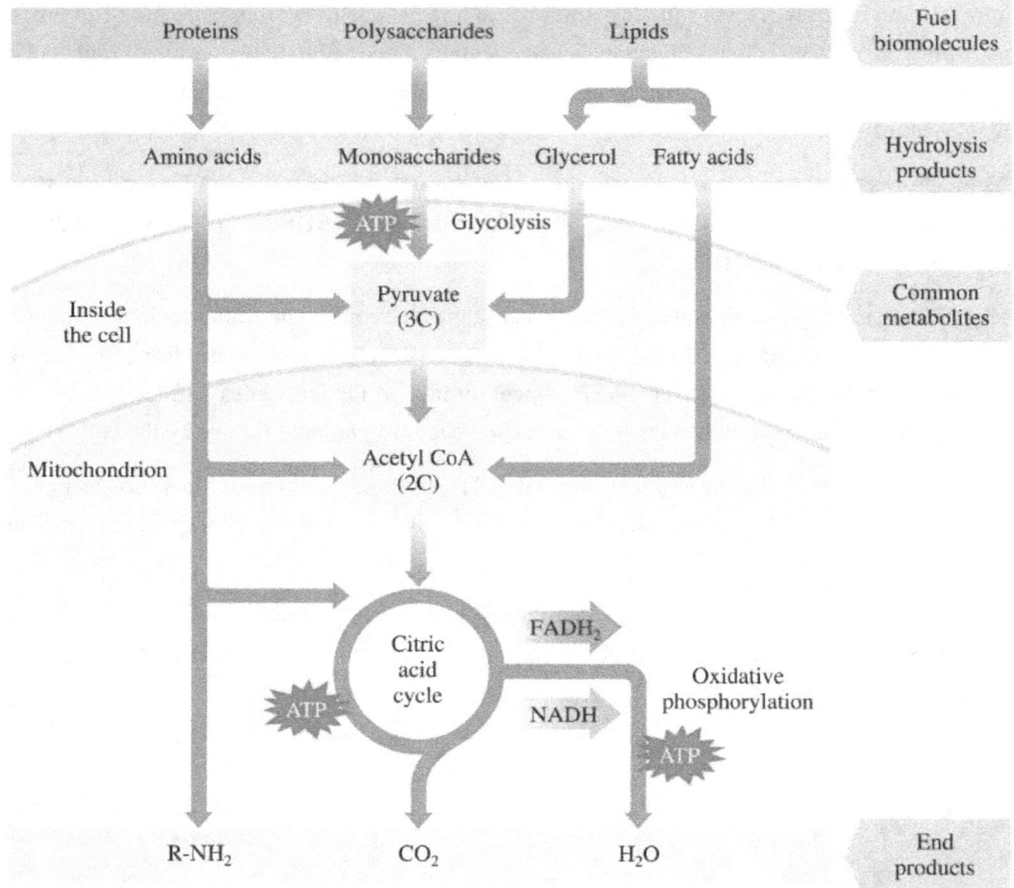

Aerobic cellular respiration showing precursor molecules used during glycolysis, citric acid cycle (Krebs) and electron transport chain

Molecules and energy produced by cellular respiration

During aerobic cellular respiration, glucose breakdown provides energy for a hydrogen ion gradient across the mitochondria's inner membrane, coupling proton flow with ATP formation.

At the end of cellular respiration, glucose is oxidized to carbon dioxide and water, and ATP is produced.

The overall chemical equation for aerobic respiration is:

$$C_6H_{12}O_6 + 6\ O_2 \rightarrow 6\ CO_2 + 6\ H_2O + \text{energy (ATP)}$$

It is the reverse of the equation for photosynthesis (e.g., plants), which uses carbon dioxide (CO_2), water, and energy from sunlight to create glucose and oxygen.

The primary purpose of cellular respiration is the production of ATP for the cell's energy needs. Glycolysis produces 2 ATP, the Krebs cycle produces 2 ATP, and the electron transport chain (i.e., oxidative phosphorylation) produces approximately 30 to 32 ATP. In total, aerobic cellular respiration results in the production of approximately 34 to 36 ATP.

The energy yield can be calculated using the energy content of glucose and ATP. Glucose contains 686 kcal/molecule, while ATP contains 7.5 kcal/molecule.

$$7.5 \times 34 = 255 \text{ kcal/mol for all ATP produced}$$

$$255 / 686 = 37.2\% \text{ energy recovered from aerobic respiration}$$

Therefore, the ~36 resulting ATP molecules represent approximately 37% of the energy in one molecule of glucose. Aerobic respiration is almost twenty times as efficient as anaerobic respiration, which has a net production of 2 ATP and an efficiency of approximately 2%.

Overall, cellular respiration is an exergonic process ($\Delta G = -686$ kcal/mole).

Regulation of oxidative phosphorylation

The movement of electrons through the electron transport chain is regulated by the proton motive force (PMF). When the magnitude of the PMF is high (i.e., a high H^+ concentration gradient), the rate of electron flow through the electron transport chain is lower. When the magnitude of the PMF is low, the rate of electron flow is greater to increase the H^+ concentration gradient.

The energy demands of the cell directly influence the regulation of oxidative phosphorylation. During resting conditions, the demand for ATP is low, and there is a low rate of proton movement from the intermembrane space through the ATP synthase in the inner mitochondrial membrane.

However, when the demand for energy is high (e.g., during vigorous activity), protons flow more quickly through the ATP synthase due to an increased concentration of ADP in the mitochondria. The level of ADP is the primary factor in determining the rate of oxidative phosphorylation. If protons are not flowing through the ATP synthases and thus allowing ADP to be phosphorylated into ATP, electrons do not flow through the cytochromes of the electron transport chain to oxygen, the terminal electron acceptor.

Oxygen is another important regulator of oxidative phosphorylation. If there is insufficient oxygen, then electrons cannot pass through the electron transport chain and NADH, and $FADH_2$ cannot be oxidized to NAD^+ and FAD. Eventually, NAD^+ and FAD are depleted, and the link reaction and Krebs cycle are suspended. By measuring oxygen consumption, the rate of the electron transport chain can be calculated (i.e., elevated oxygen consumption indicates faster electron transport). The rate of electron transport is the cellular respiratory rate.

Several compounds inhibit oxidative phosphorylation by preventing electron transport. One example is antimycin A, which inhibits cytochrome b in the coenzyme Q–cytochrome c Reductase (Complex III). Other inhibitors, such as rotenone and amytal, block the use of NADH as a substrate in NADH dehydrogenase (Complex I). The inhibitors cyanide, carbon monoxide, and azide prevent electron flow in Cytochrome c Oxidase (Complex IV).

Some compounds can inhibit oxidative phosphorylation by acting as uncoupling agents. These include 2,4-dinitrophenol (DNP) and pentachlorophenol (acidic aromatic compounds). The protons in the intermembrane space of the mitochondria pass through the membrane with the protons attached, and then the uncoupling agent releases the protein once inside the mitochondrial matrix.

This dissipates the proton motive force and allows protons to bypass the ATP synthases embedded in the membrane, preventing the production of ATP. This uncoupling of oxidative phosphorylation is a way for organisms to generate heat, (i.e., thermogenesis).

Additionally, the enzyme inhibitors oligomycin and dicyclohexylcarbodiimide can directly inhibit the ATP synthase by preventing the influx of protons.

Mitochondria and Oxidative Stress

Mitochondria for energy production

Mitochondria are organelles involved in energy production. A mitochondrion (or *cell's powerhouse*) has a double membrane with an intermembrane space between the outer and inner membrane. The *matrix* is a watery substance containing ribosomes and enzymes. Enzymes in the matrix and inner membrane catalyze the oxidation of carbohydrates, fats, and amino acids. These enzymes are vital for the link reaction and the Krebs cycle.

The inner membrane is where the electron transport chain and ATP synthase are located and where oxidative phosphorylation occurs. The space between the inner and outer membranes is a small volume space that protons are pumped into. Due to its small volume, a high concentration gradient is reached quickly, vital for chemiosmosis.

The outer membrane is a membrane that separates the contents of the mitochondrion from the rest of the cell, creating the ideal environment for cell respiration. The cristae are tubular projections of the inner membrane that increase the surface area for oxidative phosphorylation.

The parts of aerobic cellular respiration occurring in the mitochondria are 1) the link reaction (i.e., pyruvate converts to acetyl-CoA), 2) the Krebs cycle (matrix of mitochondria), and 3) oxidative phosphorylation (proton gradient within the intermembrane space of mitochondria).

During these processes, the pyruvate from glycolysis is broken down to CO_2 and H_2O, which is why aerobic respiration is considered as glucose breakdown.

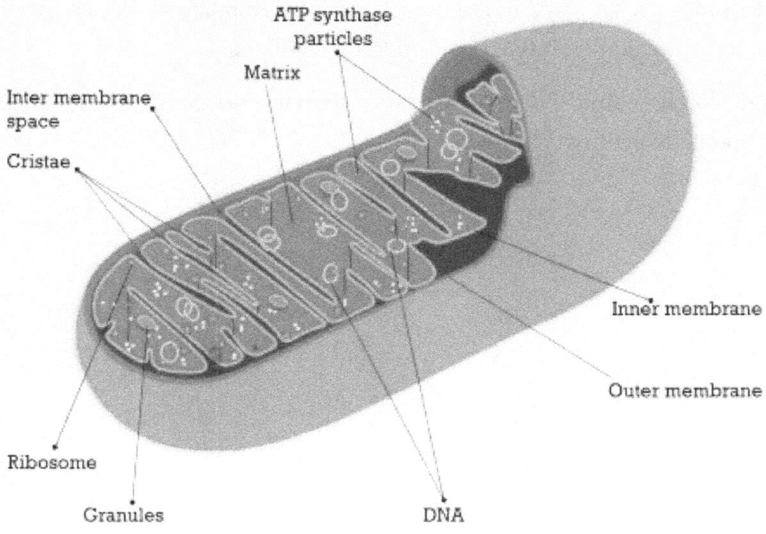

Mitochondrion with cellular compartments and select biomolecules

CO_2 and ATP are transported from the mitochondria to the cytoplasm. CO_2 enters the bloodstream for transport to the lungs for expiration. H_2O remains or enters the blood for excretion by the kidneys.

Apoptosis is the programmed cell death that occurs in multicellular organisms. It occurs in both developing tissue (e.g., in embryonic development where parts of tissues are no longer needed) and adult tissue (where cell death balances cell division). When a cell dies by apoptosis, its cytoskeleton collapses, its nucleus fragments, and chromatin irreversibly condenses.

After apoptosis, specific neighboring cells (e.g., macrophages) recognize alterations in the surface of the dead cell and phagocytize the cell before its contents are leaked, allowing for efficient recycling of the cell's biomolecules.

The machinery for apoptosis depends on an intracellular proteolytic cascade (which involves the breaking of peptide bonds within proteins). Target proteins (both enzymes and structural proteins) are cleaved by proteases, leading to the activation of more proteases.

Eventually, this leads to the degradation of all the cell's organelles. However, before this cascade begins, signals must cause the initiation of the apoptosis pathway. Apoptosis usually begins in response to stress; nuclear receptors recognize stress factors, including nutrient deprivation, heat, viral infection, radiation, and lack of oxygen.

These stress factors lead to the release of intracellular apoptotic signals that cause proteins to initiate the apoptotic pathway.

Mitochondria are the target of some apoptotic proteins.

These proteins may create pores in the mitochondrial membrane and cause swelling, or they may change the permeability of the mitochondrial membrane.

When the mitochondrial membrane is made more permeable, apoptotic effectors leak out and activate the apoptotic pathway's proteins.

Free radicals and oxidative stress

Oxidative stress is an imbalance between reactive oxygen species (i.e., free radicals) and the organism's ability to detoxify harmful substances.

Free radicals are atoms or molecules with an unpaired electron (single dot for an unpaired electron in the molecule). They have an odd number of electrons.

Free radicals, such as the hydroxyl radical (HO•), are highly reactive, and if they react with important cellular components such as the cell membrane or DNA, the cell can be severely damaged.

The extent of the damage can be extensive because when a free radical abstracts an electron from a cell component, the cell component must take an electron from another cell component, leading to a chain of free radical reactions.

The creation of free radicals disrupts cellular signaling if the original molecule were a cellular messenger.

Oxidative stress is involved in many diseases, including cancer, Alzheimer's disease, and Parkinson's disease. However, oxidative stress can be beneficial (e.g., when it attacks pathogens rather than the host's cells).

Antioxidants are molecules that donate electrons to free radicals without becoming destabilized, and they are the primary method by which the reactive oxygen species are counteracted.

Oxidative stress is caused by a decrease in antioxidant levels, increased free radical levels, or both.

Sometimes, oxidative stress can trigger apoptosis or necrosis (cell injury leading to premature cell death).

Notes for active learning

Principles of Metabolic Regulation

Homeostasis as a dynamic steady state

Metabolism refers to the biochemical reactions occurring in living organisms' cells, including the reactions that allow the cell to grow, reproduce, and function.

Because organisms are usually in environments continually changing, metabolism must be tightly regulated so that the cell's internal conditions remain constant.

This is *homeostasis* and establishes a "dynamic steady-state."

Homeostatic processes act at multiple levels, from the cell or the tissue to the whole organism. Examples include the regulation of pH, blood glucose, and internal body temperature.

Catabolic (breaking down) and anabolic (building up) pathways are extensively regulated.

During the metabolic reactions in steady-state conditions, the substrate is converted to the product as efficiently as possible.

Regulation of glycolysis and gluconeogenesis

Glycolysis, the breakdown of glucose into two pyruvate molecules, is universal to biological organisms. Glycolysis is tightly regulated. The three most important steps in the pathway—the reactions catalyzed by phosphofructokinase (phosphorylation of fructose-6-phosphate), hexokinase (phosphorylation of glucose), and pyruvate kinase (phosphate transfer from phosphoenolpyruvate to ADP)—are the most closely regulated because these three steps are essentially irreversible due to a large $-\Delta G$. T

he production of these three enzymes is regulated by hormones that control the rate of transcription (DNA→mRNA). For more immediate action, allosteric effectors bind to these enzymes reversibly, or the enzymes can be covalently modified to inhibit activity permanently.

Phosphofructokinase is the most important enzyme for glycolysis regulation. High ATP levels allosterically inhibit phosphofructokinase, meaning that the ATP binds to an allosteric site distinct from the catalytic active site. ATP inhibits the key reaction in glycolysis because if there are high levels of ATP, there is no need for more to be produced.

Citrate, one of the early intermediates in the Krebs cycle, inhibits phosphofructokinase. It does this by enhancing ATP's inhibitory effect. When there are high levels of citrate, there is no need for additional glucose breakdown.

Some compounds activate phosphofructokinase. For example, the activator AMP reverses ATP's inhibitory action and allows the catalytic activity of phosphofructokinase to resume. Fructose 2,6-bisphosphate, another allosteric activator, reduces ATP's inhibitory effect and increases the enzyme's affinity for fructose-6-phosphate.

The concentration of fructose 2,6-bisphosphate is controlled by two enzymes: phosphofructokinase 2 (different than phosphofructokinase) and fructose bisphosphate 2.

These *bifunctional enzymes* are present in a single polypeptide chain. The activity of this bifunctional enzyme is controlled by the phosphorylation of a serine residue in the polypeptide chain.

The phosphorylation of this residue, which occurs when glucose levels are low, leads to a reduction in fructose 2,6-biphosphate production and thus a reduction in the rate of glycolysis, thereby conserving glucose.

However, when glucose is abundant, dephosphorylation of the serine residue occurs, leading to an increase in fructose 2,6-biphosphate production and thus an increase in the rate of glycolysis. This regulation modulates the levels of glucose so it is available when needed without being degraded.

Hexokinase, another important regulatory enzyme in glycolysis, is inhibited by its product, glucose-6-phosphate. Since glucose-6-phosphate is in equilibrium with fructose 6-phosphate (the reactant of phosphofructokinase), the inhibition of phosphofructokinase results in the inhibition of hexokinase. This further suggests that phosphofructokinase is the key enzyme in glycolysis regulation.

Pyruvate kinase controls outflow from the glycolysis pathway and is thus important in regulation. It is activated by fructose 1,6-bisphosphate as an example of *feedforward stimulation* since fructose 1,6-bisphosphate is an earlier intermediate in the pathway that enables pyruvate kinase to keep up with the incoming flux of intermediates. Pyruvate kinase is allosterically inhibited by ATP (which signals that energy is abundant) and alanine (which signals that building blocks are abundant).

Transcription of the enzymes used in glycolysis is controlled by hormones, which allows for coordination between different organs and tissues. These hormones regulate the rate of gluconeogenesis (glucose from non-carbohydrate substrates).

Gluconeogenesis and glycolysis are "reciprocally regulated," meaning that one pathway is active while the other is inactive. This prevents the occurrence of a *futile cycle*, where ATP is hydrolyzed without useful metabolic work, resulting in chemical energy dissipating as heat and being wasted.

As discussed earlier, gluconeogenesis uses glucose-6-phosphatase, fructose-1,6-bisphosphatase and PEP carboxykinase/pyruvate carboxylase, replacing three strongly endergonic reactions with reactions that are more exergonic (and therefore more favorable).

These replacements allow for the reciprocal regulation of these two pathways; i.e., the same compounds that activate one enzyme in glycolysis may inhibit the replacement enzyme in gluconeogenesis.

For example, phosphofructokinase in glycolysis is activated by AMP and inhibited by ATP and citrate; fructose-1,6-bisphosphatase, however, is inhibited by AMP and activated by ATP and citrate. Since gluconeogenesis consumes ATP while glycolysis produces ATP, high levels of ATP promote gluconeogenesis (by activation of fructose-1,6,-bisphosphate) and reduce the rate of glycolysis (by inhibition of phosphofructokinase).

Acetyl-CoA is an important activator of gluconeogenesis. It activates pyruvate carboxylase to catalyze the first reaction in gluconeogenesis. Conversely, it inhibits the enzyme pyruvate kinase in glycolysis, demonstrating the reciprocal control of the two pathways.

Metabolism of glycogen

Glycogen is a large, branched polysaccharide made of glucose residues, linked primarily by α-1,4-glycosidic bonds, although one of every ten bonds is an α-1,6-glycosidic bond.

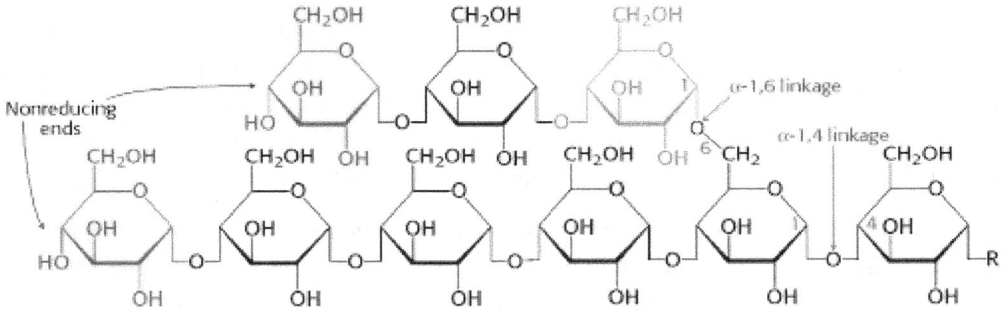

Glycogen with α-1,4-glycosidic and α-1,6-glycosidic bonds

It is the main form of stored glucose in the body (mostly stored in the liver and skeletal muscles) because it can easily be broken down when energy is needed, allowing monomers of glucose to be released when levels are low.

When glycogen is metabolized in the liver, it can be slowly released into the bloodstream (e.g., between meals). When it is metabolized in muscle, it can provide a quick burst of glucose for energy production of ATP for aerobic or anaerobic activity. Oxygen is not required for glycogen breakdown.

Glycogenolysis is the breakdown of glycogen in the following reaction:

$$\text{glycogen}_{(n\text{ residues})} + P_i \rightleftarrows \text{glycogen}_{(n-1\text{ residues})} + \text{glucose 1-phosphate}$$

There are three main steps in glycogen metabolism. The first step is the release of a glucose-1-phosphate molecule from glycogen, which is catalyzed by glycogen phosphorylase. It is a phosphorolysis reaction (breakdown by the addition of a phosphate molecule).

Glycogen phosphorylase only acts on non-reducing ends of a glycogen polymer that are five or more glucose residues away from a branch point (see figure above).

In the second step, the glycogen is remodeled for further breakdown of the polysaccharide. When there are only four glucose residues before a branching point, a glycogen debranching enzyme (a type of transferase) transfers three of the remaining four units of glucose (a trisaccharide) from the 1,6 branch to an adjacent 1,4 branch.

This exposes the 1,6 branching point for hydrolysis by α-1,6-glucosidase. The final glucose molecule is removed, and the branch is eliminated, thus allowing the phosphorolysis of glycogen by glycogen phosphorylase to continue.

The third step of glycogen metabolism is the conversion of glucose-1-phosphate to glucose-6-phosphate by the enzyme phosphoglucomutase, which uses a phosphoserine to remove a phosphoryl group.

At this point, there are three possibilities for the glucose-6-phosphate molecule resulting from the glycogen breakdown: (1) enter the glycolysis pathway for the production of ATP, (2) enter the pentose phosphate pathway for the production of NADPH and ribose derivatives, or (3) become dephosphorylated and converted into free glucose for release into the bloodstream by the enzyme glucose 6-phosphatase (i.e., the final step in gluconeogenesis).

Allosteric and hormonal control of glycogen synthesis and degradation

The regulation of glycogen anabolism and catabolism is complex. If both pathways were to occur at the same time, it would be a futile cycle, so the simultaneous action of these pathways must be avoided. Like glucose synthesis or breakdown, glycogen synthesis and breakdown are reciprocally regulated.

Many of the enzymes involved in these processes are controlled allosterically, and they respond to metabolites that signal the cell's energy needs. The enzyme activity is adjusted so that the demands for energy are met. Glycogen phosphorylase is inhibited by high-energy molecules such as ATP, glucose-6-phosphate, and glucose, and is activated by low-energy molecules such as AMP.

In addition to allosteric control, hormones regulate glycogen synthesis and breakdown. Hormones trigger a cAMP (cyclic AMP) cascade, which is a signaling pathway that can act through the enzyme protein kinase A (PKA). PKA can activate phosphorylase kinase and deactivate glycogen synthase (through the addition of a phosphoryl group), which prevents glycogen synthesis and catabolism from occurring simultaneously.

Conversely, protein phosphatase 1 reverses the regulatory effects of PKA. Insulin activates protein phosphatase 1, which stimulates the synthesis of glycogen. Hormonally-stimulated cascades allow the rate of glycogen synthesis or catabolism to be adjusted by the needs of both cells and the entire organism.

Analysis of metabolic control

Metabolic *regulation* is the changes that signaling molecules cause on enzyme activity for catalyzing reactions in metabolic pathways.

Metabolic *control* is how changes in enzyme activity control flux (i.e., the overall rate) of the pathway.

These two concepts are intricately linked with multiple levels of metabolic control.

Intrinsic and extrinsic control of metabolism

There are multiple levels of metabolic control. *Intrinsic control* occurs when the reactions in metabolic pathways are self-regulated by responding to changes in the levels of products and substrates.

For example, the product of glycogen phosphorylase is glucose-6-phosphate, and it is inhibited by glucose-6-phosphate (feedback inhibition or negative feedback). Positive feedback occurs when a reaction's product amplifies the reaction rate (e.g., blood clotting).

Extrinsic control occurs when a cell in a multicellular organism alters its metabolism according to a signal sent from another cell. These signals are usually hormones or growth factors.

They are detected by receptors on the affected cell's surface, leading to the transmission of signals within the cell (second messenger system). An example of extrinsic control is the hormone-triggered phosphorylation cascade that reduces the rate of glycogen synthesis (described earlier).

Extrinsic control systems allow for the maintenance of homeostasis at the whole-organism level.

There are three components to homeostatic control mechanisms that enable this to happen.

The *receptor* is the first component, which senses environmental stimuli.

The receptor sends a signal to the second component, the *control center* (e.g., brain).

The control center sends a signal to the third component, the *effector*, which responds to the stimuli.

Notes for active learning

Metabolism of Fatty Acids

Fatty acids as an energy source

When glucose supply is low, the body uses other energy sources: other carbohydrates, fats, and then proteins. First, these molecules are converted to glucose or glucose intermediates; they are degraded in glycolysis or by the Krebs cycle.

Monosaccharides and amino acids travel directly through the bloodstream to the cells for absorption. Triglycerides are packaged into lipoproteins (chylomicrons) for delivery.

Description of fatty acids

Fatty acids are a family of molecules classified as lipids, which are usually ingested and stored as triglycerides. A *triglyceride* (triacylglycerol) consists of a three-carbon glycerol backbone with three fatty acids connected to it, as displayed below.

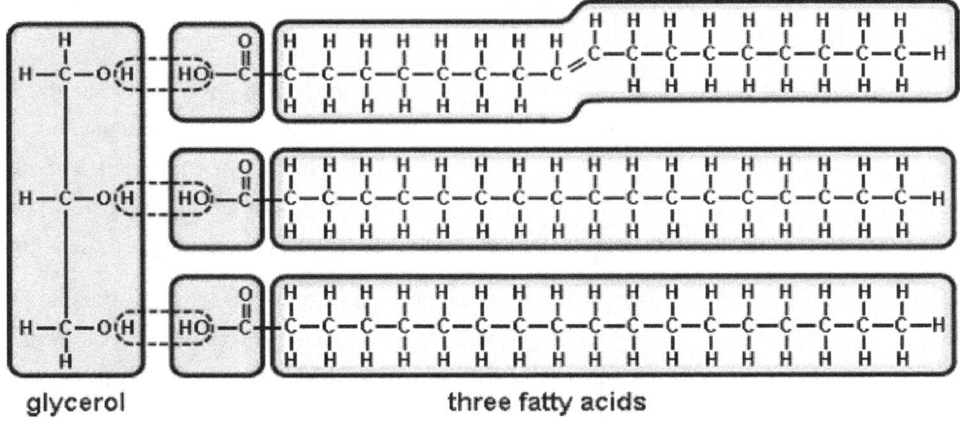

Glycerol and three fatty acids are each joined by a phosphodiester bond

Saturated fats have hydrogens that occupy all possible bonds and have only single bonds. *Unsaturated* fats do not have hydrogen atoms at all positions, and there is at least one double bond.

Fatty acids are an essential energy source because they are reduced and anhydrous (hydrophobic), which allows for a higher energy yield. Carbohydrates are hydrated, so the amount of energy stored per unit mass is much lower than fatty acids.

For this reason, it is ideal for an organism to store energy as fat in the adipose tissue when a large quantity of energy needs to be kept in reserve for later use (e.g., hibernating bear).

trans-Oleic acid

cis-Oleic acid

Cholesterol is a four fused ring structure

Cis and trans unsaturated fatty acids

Digestion, transport and storage of fats

The pancreatic enzyme lipase breaks down triglycerides into free fatty acids and monoglycerides.

The pancreatic enzyme lipase breaks down triglycerides into free fatty acids and monoglycerides (3-carbon glycerol and fatty acid) by hydrolyzing the ester bond, because triglycerides cannot be absorbed by the duodenum (first segment of the small intestine). Pancreatic lipase forms a complex with the protein colipase, which is essential for its activity, and this complex works at a water-fat interface because dietary fats are nonpolar molecules.

Bile is excreted from the gallbladder into the stomach to assist in digestion. Bile contains *bile salts*, which are *amphipathic* (i.e., having both hydrophilic and hydrophobic parts) to orient their nonpolar face toward the dietary fats and their polar face toward the water, forming *micelles*, which are spherical lipid droplets.

Emulsification is the breaking of larger nonpolar globules into micelles. The micelles move the dietary fats closer to the intestinal cell wall so that cholesterol can be absorbed across the intestinal wall and triglycerides can be hydrolyzed.

Once across the intestinal wall, free fatty acids and monoglycerides are reassembled as triglycerides, while the cholesterol is linked to another free fatty acid, forming a cholesterol ester.

These are repackaged as chylomicrons, lipoproteins that transport triglycerides in the bloodstream and the tissues, used for energy production or stored. The liver is an essential organ for fatty acid metabolism.

Adipocytes are cells where most body fat is stored, as the entire cytoplasm is filled with a single fat droplet. Adipocytes synthesize and store triglycerides during food uptake and cluster to form adipose tissue beneath the skin and around internal organs.

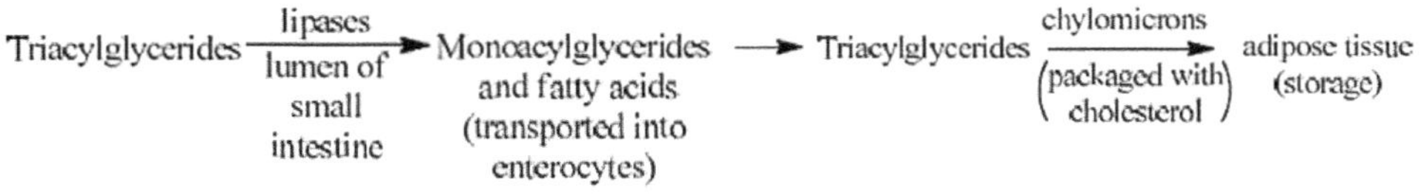

$$\text{Triacylglycerides} \xrightarrow[\substack{\text{lumen of}\\ \text{small}\\ \text{intestine}}]{\text{lipases}} \substack{\text{Monoacylglycerides}\\ \text{and fatty acids}\\ \text{(transported into}\\ \text{enterocytes)}} \longrightarrow \text{Triacylglycerides} \xrightarrow[\substack{\text{(packaged with}\\ \text{cholesterol)}}]{\text{chylomicrons}} \substack{\text{adipose tissue}\\ \text{(storage)}}$$

Beta oxidation of fatty acids

Fatty acids are degraded to produce ATP when glucose supplies are low.

For fatty acids to undergo catabolism, they first must be activated. Catalysis by the enzyme fatty acyl-CoA synthase links fatty acids to coenzyme A, forming activated fatty acyl-CoA. Then, the fatty acyl-CoA is transported to the mitochondrial matrix with the help of the enzymes: carnitine acyltransferase I (which conjugates fatty acyl-CoA to carnitine), carnitine-acylcarnitine translocase (which shuttles carnitine and acylcarnitine inside the mitochondria), and carnitine acyltransferase II (which liberates the carnitine and turns the molecule back into fatty acyl-CoA). However, this applies to long-chain fatty acids; short-chain fatty acids can diffuse directly into the mitochondrial membrane with no need for the carnitine carrier system.

The molecule undergoes β oxidation (beta carbon of the fatty acid is oxidized to a carbonyl group).

In this process, fatty acyl-CoA undergoes oxidation by repeatedly cleaving two-carbon molecules, each time producing a new fatty acyl-CoA that is two carbons shorter, along with one molecule of acetyl-CoA.

β oxidation works best for even-numbered saturated fatty acids due to the repetitive cleaving of two carbons from the chain. Saturated fatty acids produce one NADH and one $FADH_2$ for every two carbons.

The acetyl-CoA produced during β oxidation enters the Krebs cycle, and the typical progression of aerobic cellular respiration occurs. A short eight-carbon fatty acid can produce four acetyl-CoA.

Each acetyl-CoA yields 12 ATP (3 NADP, 1 $FADH_2$, and 1 ATP).

Therefore, this eight-carbon fatty acid nets 48 ATP, and fat with three chains of this length produces 144 ATP, illustrating why fats are an excellent source of energy.

Fats provide 9 calories per gram, while carbohydrates and proteins provide 4 calories per gram each.

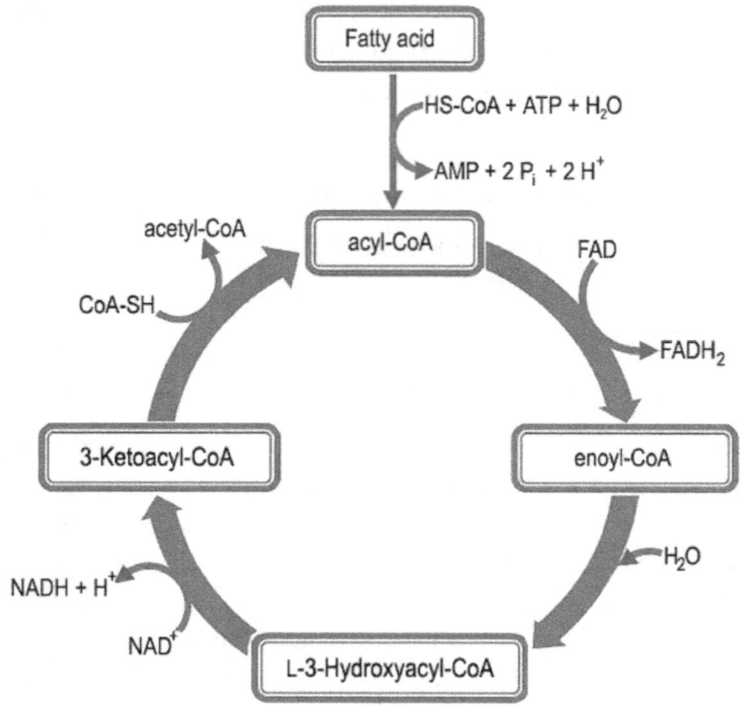

Each turn of β oxidation produces one NADH and one FADH₂,
used in the electron transport chain to produce ATP

For odd-numbered saturated fatty acids in the lipids of some marine organisms and plants, the products of β oxidation are propionyl-CoA and acetyl-CoA. With the help of three enzymes and the vitamins cobalamin and biotin, the propionyl-CoA undergoes carboxylation and molecular rearrangement to form succinyl-CoA. The succinyl-CoA then enters the Krebs cycle.

Unsaturated fatty acids require additional enzymes for β oxidation because the double bonds in the fatty acid are usually in a *cis* configuration, which causes steric hindrance.

First, β oxidation of the unsaturated fatty acid occurs normally, but when the presence of a *cis* bond interferes with the ability of acyl-CoA dehydrogenase or enoyl-CoA hydratase to catalyze the necessary reaction, the *cis* bond must be converted into a *trans* bond.

Odd-numbered *cis* bonds are transformed by enoyl-CoA isomerase, which changes the *cis* bond to a *trans* bond. Even-numbered *cis* bonds are transformed by 2,4-dienoyl-CoA reductase, which creates an odd-numbered bond, and the enoyl CoA isomerase can then act.

Unsaturated fatty acids yield one less FADH₂ per double bond compared to saturated fatty acids because they are already partially oxidized, thus reducing total ATP production.

The glycerol that was originally part of the stored triglyceride is converted to glyceraldehyde phosphate and enters glycolysis or gluconeogenesis (depending on the cell's needs at the time).

However, it must be converted to glyceraldehyde-3-phosphate (PGAL) before it can enter either of these pathways.

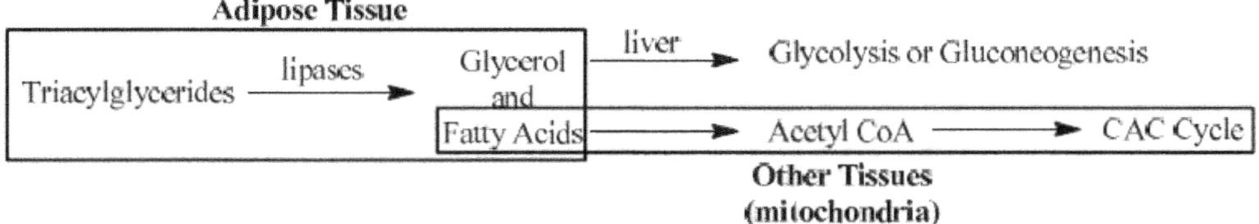

Ketogenesis

Sometimes there is an insufficient supply of glucose in the body (e.g., during fasting). Once the cellular carbohydrate stores have been depleted, the organism must find another way to obtain energy.

Ketogenesis is the production of ketone derivatives of acetyl-CoA groups as another form of fatty acid breakdown and occurs in the mitochondria of the liver.

The oxidation of large amounts of fatty acids can cause acetyl-CoA to accumulate in the liver.

If the quantities of acetyl-CoA are high, it may exceed the Krebs cycle's processing capacity. This can occur when excess deaminated ketogenic amino acids are degraded, causing acetyl-CoA buildup. In cases of high acetyl-CoA accumulation, the ketogenesis pathway is initiated.

Oxidation of fatty acids

In ketogenesis, the two-carbon acetyl units condense in the liver, forming the following four carbon ketone molecules: β-hydroxybutyrate, acetoacetate, and acetone. These are collectively referred to as *ketone bodies.*

These ketone bodies are transferred from the liver to the heart, brain, and muscles. In these organs, β-hydroxybutyrate and acetoacetate are reconverted to acetyl-CoA for entry into the Krebs cycle.

Acetone is converted to pyruvate, lactate, and acetate or excreted as waste if it is not used quickly.

Ketosis is when an excessive amount of ketone bodies is present. In this metabolic state, energy comes from ketone bodies in the blood. It occurs during fasting or low-carbohydrate diets and is "fat-burning mode."

If the body fails to regulate ketone production properly, the excessive formation of ketone bodies can cause *ketoacidosis* or *metabolic acidosis* since acetoacetate and *β*-hydroxybutyrate are acidic. In this condition, blood pH is drastically decreased and can sometimes be fatal.

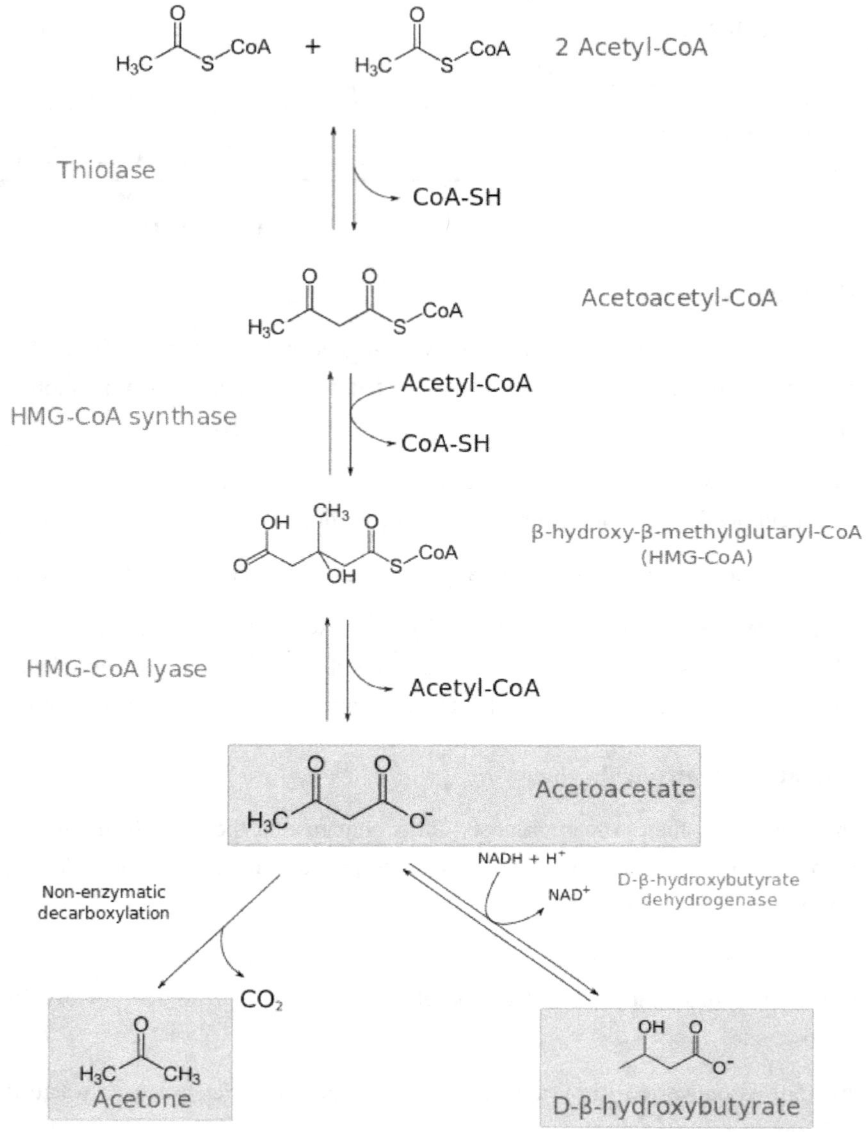

Conversion of acetyl-CoA to ketones (acetone) and β-hydroxybutyrate

Lipogenesis and essential fatty acids

When there are excess dietary carbohydrates, the molecules must be converted to fat for storage.

The anabolism of fats primarily occurs in the cytoplasm and the endoplasmic reticulum of liver cells, in contrast with the fatty acid breakdown, which occurs in the mitochondria.

Lipogenesis is when fats are produced from acetyl-CoA and malonyl-CoA precursors by fatty acid synthases that polymerize and reduce acetyl groups.

Lipogenesis is stimulated by insulin. The synthesized fatty acids are esterified with glycerol to create triglycerides, and the triglycerides are packaged into lipoproteins and secreted from the liver.

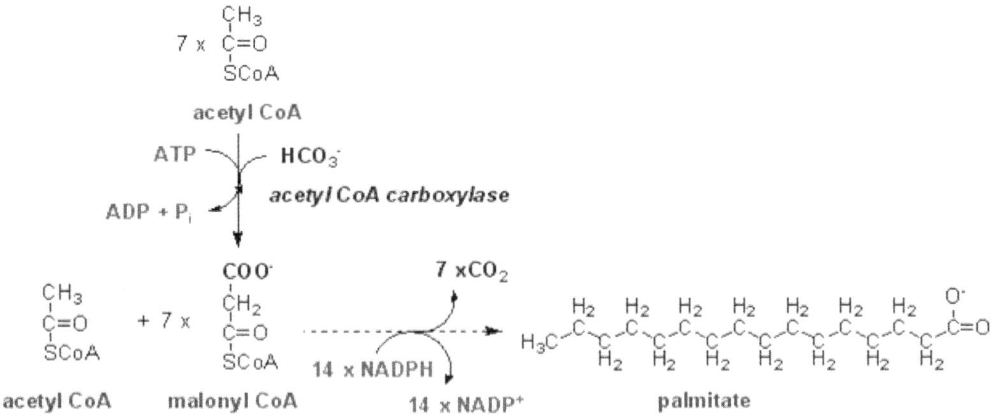

Synthesis of palmitate, the primary fatty acid synthesized in the human body

As seen in the diagram above, ATP and NADPH are required for fatty acid synthesis. During the process, these molecules are oxidized to ADP and $NADP^+$, respectively.

To synthesize an unsaturated fatty acid, a desaturation reaction introduces a double bond into the fatty acyl chain (usually requiring a desaturase enzyme).

Essential fatty acids cannot be synthesized in mammalian tissue and are required in the diet.

The essential fatty acids are linolenic acid (omega-3 fatty acid) and linoleic acid (omega-6 fatty acid). If these are not consumed in the diet causing deficiency of essential fatty acids, health problems may develop.

Non-template synthesis: biosynthesis of lipids and polysaccharides

Lipids are a group of molecules that include triglycerides, fat-soluble vitamins, waxes, sterols, and phospholipids. Polysaccharides (carbohydrate polymers) are synthesized by non-template *de novo* synthesis.

For *de novo* synthesis, there is no template as there is in the synthesis of nucleic acids or proteins; rather, the synthesis of carbohydrate polymers is based solely on gene expression and enzyme specificity.

Additionally, other molecules are synthesized *de novo*. Terpenes, a class of lipids that exists in all organisms are created by the assembly and modification of isoprene units. These molecules can then be used to create sterols (e.g., cholesterol).

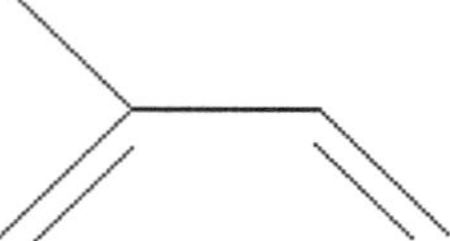

Isoprene as a component of carotenoids, steroids, etc.

Glycogen is a polysaccharide that is synthesized *de novo*, constructed by joining units of uracil-diphosphate-glucose by the enzyme glycogenin, which acts as a primer. Glycogen synthase adds glucose monomers to the existing chain, and the glycogen branching enzyme creates branches.

Dietary protein and cellular metabolism of amino groups

When carbohydrates and fats are unavailable, proteins are used as an energy source. Proteins are the least desirable energy source for the body because a large amount of energy is required for protein breakdown.

Proteolysis (protein catabolism) requires proteases to break the peptide bonds of proteins via hydrolysis to release smaller polypeptides or amino acids.

Amino acids can provide intermediates for a variety of other molecules.

Most amino acids are deaminated (the amino group is removed) in the liver and are converted into pyruvate, acetyl-CoA, or other Krebs cycle intermediates. Depending on the amino acid, these intermediates enter cellular respiration at various points.

From dietary intake, protein digestion begins in the stomach, where proteins are denatured (unfolded) by the acidic digestive juices. Digestive enzymes like pepsin, trypsin, and chymotrypsin hydrolyze covalent peptide bonds. Amino acids are absorbed in the small intestine into the bloodstream for delivery to the tissues.

Keto acids for ATP synthesis

Amino acids can produce ATP when other fuel supplies are low, and the cell does not require other nitrogen-containing compounds.

To produce ATP, the amino acid group must be removed by oxidative deamination to produce a keto acid and NH_3 or transferred to a keto acid by *transamination* (transfer of an amino acid group to an organic acid).

The keto acid enters the glycolytic pathway or the synthetic pathways for glucose and fat. The amino group's nitrogen is used to synthesize critical nitrogen-containing molecules such as purines and pyrimidines.

The toxic NH_3 (ammonia) passes into the bloodstream through the plasma membrane. It is transported to the liver, linked with CO_2 to form, in mammals, the relatively non-toxic urea, excreted by the kidneys. In fish, insects, and birds, ammonia is converted to uric acid rather than urea.

Amino acids may replenish the intermediates in the Krebs cycle.

Three-carbon amino acids (e.g., alanine) enter the pathways as pyruvate.

Four-carbon amino acids (e.g., aspartate) are converted to oxaloacetate.

Five-carbon amino acids (e.g., valine) are converted to α-ketoglutarate.

Some amino acids can enter at more than one point, depending on cellular requirements.

Essential amino acids

Plants synthesize amino acids as needed; animals lack some enzymes needed to make some amino acids.

Humans synthesize eleven of the twenty necessary amino acids.

The diet must provide the remaining nine amino acids as *essential amino acids*.

The total free amino acid pool in the body is derived from:

      1) ingested protein degraded to amino acids during digestion,

      2) synthesis of non-essential amino acids from keto acids, and

      3) the breakdown of body proteins.

The amino acids in these pools are used for protein biosynthesis.

Notes for active learning

Hormonal Regulation and Integration of Metabolism

Higher-level integration of hormone functions

Hormones are signaling molecules produced by glands in multicellular organisms and travel in the bloodstream to their target cell, organ, or tissue. These molecules are essential for regulating cell metabolism, as evidenced by several hormonal regulation examples in this chapter.

There are three main categories of hormones based on their chemical structure: peptide hormones, steroid hormones, and modified amino acid hormones.

Peptide hormones target receptors

Peptide hormones range from three amino acids to folded proteins with subunit structures. When they have longer amino acid chain lengths, they are protein hormones. These hormones, which include insulin and parathyroid hormone, are initially produced in endocrine tissue as large products and undergo extensive processing before they are stored in vesicles or granules in preparation for release.

Peptide hormones have short half-lives in the bloodstream, so they are released at the precise moment they are needed. The response is rapid because these hormones are stored in large quantities. When released, peptide hormones target protein and glycoprotein receptors embedded in cell membranes.

The binding of the hormone to the receptor on the cell's surface can initiate a signal transduction cascade within the cell, which may have many effects. These include increasing uptake of a molecule, phosphorylation or dephosphorylation of target molecules within the cell, triggering secretion, or activating mitosis.

Steroid hormones affect gene transcription

Steroid hormones, such as glucocorticoids and progestins, are synthesized from cholesterol, usually in the adrenal glands and the gonads.

The synthesis of these hormones occurs in the mitochondria and smooth endoplasmic reticulum, where a series of enzymatic reactions convert cholesterol into the specified steroid hormone.

Unlike polypeptide hormones, steroid hormones are not stored in large quantities. Most hydrophobic steroid hormones circulating in the bloodstream are bound to carrier proteins, and the unbound steroid hormones are the steroid molecules entering the target cells.

Steroid hormones easily pass through cell membranes because they are hydrophobic and fat-soluble. Inside the cell, they bind to intracellular receptors, causing a conformational change in a transcriptional complex.

This increases or decreases the transcription rate (DNA→RNA) of genes and consequently affects the rate at which specific proteins are produced.

An example is glucocorticoids that bind to the glucocorticoid receptor complex, which inhibit pro-inflammatory transcription factors, thus having an anti-inflammatory effect.

Modified amino acids affect cell metabolism

Modified amino acid hormones are the third category of hormones. Examples are norepinephrine, epinephrine, and the thyroid hormones thyroxine (T_4) and triiodothyronine (T_3). The chemical modifications creating modified amino acid hormones include methylation, decarboxylation, or hydroxylation.

Modified amino acid hormones may have a long half-life like steroid hormones or a short half-life like peptide hormones.

Depending on the hormone, they may bind to receptors on the cell-surface (like peptide hormone) or to intracellular receptors (like steroid hormone).

They are a group of hormones with significant effects on cell metabolism.

Tissue-specific metabolism

Each tissue of the body has a specialized function reflected in its anatomy, biomolecules, and metabolic activity. Skeletal muscle tissue, for instance, allows for directed motion, while adipose tissue stores and releases energy in the form of fat.

The brain pumps ions across the neurons' plasma membranes to produce electrical signals. The liver processes and converts nutrients that include carbohydrates, fats and proteins, and distributes them to the appropriate locations.

The liver detoxifies foreign compounds like drugs, preservatives, and food additives. The liver is responsible for synthesizing fats. The process of lipogenesis, where acetyl-CoA is converted to fatty acids through the addition of two-carbon units, occurs in the cytoplasm.

Most of the enzymes involved in fatty acid synthesis are arranged into a multienzyme complex as fatty acid synthetase. In animals and humans, fatty acids are usually stored in adipose tissue and the liver as a triglyceride.

Insulin regulates blood sugar levels

The hormone insulin is an indicator of blood sugar levels. The level of insulin increases as blood glucose levels increase. Insulin increases the rate of storage pathways (e.g., lipogenesis) by stimulating the pyruvate dehydrogenase complex (PDC), leading to the formation of acetyl-CoA and acetyl-CoA carboxylase (ACC), which forms malonyl-CoA from acetyl-CoA.

As insulin levels increase, the malonyl-CoA levels increase; this is important because the synthesis of malonyl-CoA is the first committed step in the fatty acid synthesis.

Insulin affects ACC in a way similar to PDC; dephosphorylation leads to the enzyme's activation.

Glucagon for glycogenolysis

Glucagon has an agonistic effect and increases phosphorylation, inhibiting ACC and slowing fat synthesis.

Muscle tissue is involved in the mechanical movements of the body. Additionally, muscle tissue, which consumes large quantities of oxygen, is an ATP generator.

Muscular activity releases epinephrine from the adrenal medulla; epinephrine binds to a receptor on the muscle cell membrane and initiates adenyl cyclase in the membrane, triggering the cAMP cascade (described previously).

Epinephrine is involved in glycogen breakdown in the muscles and occasionally in the liver.

The protein kinase, activated by cAMP, causes phosphorylations on a series of enzymes to produce glucose-1-phosphate, later converted to glucose-6-phosphate.

Simultaneously, enzymes activate glycogen breakdown (glycogenolysis), while other enzymes such as glycogen synthetase are inactivated to inhibit glycogenesis.

The two hormones involved in the control of glycogenolysis are the peptide hormone glucagon (released from the pancreas when blood glucose levels are low) and the catecholamine epinephrine (released from the adrenal glands in response to a threat or stress).

Glucagon and epinephrine activate enzymes to initiate glycogen phosphorylase to start glycogenolysis, and they inhibit glycogen synthetase.

The stimulation by epinephrine and other hormones results in the production of glucose-6-phosphate, which may proceed through glycolysis; this can occur in a muscle cell. However, if it happened to a liver cell stimulated by glucagon, glucose-6-phosphate would be converted to glucose and released into the bloodstream. During muscle contractions, ATP is constantly being used to supply energy.

The *Cori cycle* involves recycling lactic acid in the liver and muscles. In this cycle, lactate goes through gluconeogenesis, producing glucose. If muscular activity continues in the body, the amount of oxygen at the end of the electron transport chain becomes the limiting factor, and soon the oxygen supply is completely exhausted. When this occurs, the Krebs cycle is inhibited, and pyruvate starts to accumulate. Epinephrine stimulates the enzymes involved in glycogenolysis, and more glucose-6-phosphate is produced.

When the cells become anaerobic, the process of glycolysis continues as pyruvic acid is converted to lactic acid, which shifts part of the metabolic burden to the liver. The formation of lactic acid requires NADH from glycolysis and produces NAD^+ to allow the glycolysis pathway to perpetuate.

If, however, strenuous muscle activity has stopped, glucose activity can be utilized to rebuild supplies of glycogen through glycogenesis. During this time, the oxygen debt is eliminated as O_2 is replenished. For lactic acid to be converted to glucose, some of it must be transformed into pyruvic acid and then to acetyl-CoA. The Krebs cycle and electron transport chain is initiated, which produces ATP to give energy for glycogenesis of the remaining lactic acid to glucose.

Hormonal regulation of biochemical molecules

Thyroid hormones are essential for controlling metabolism and play a permissive role in developing and maintaining the nervous system. Thyroid hormones are essential determinants of basal metabolic rate (BMR). They can increase BMR by increasing oxygen consumption, and heat-production in most body tissues termed a *calorigenic effect.*

Blood glucose concentration levels fluctuate throughout the day. Homeostasis targets plasma blood glucose levels of around 80 to 120 mg/dl. Through negative feedback, specific target organs affect the rate at which glucose is taken up from or released into the blood.

When glucose levels are too high (above the homeostasis set point), β cells in the pancreatic islets produce insulin, which stimulates muscle cells and liver cells to take up glucose from the blood and convert it into glycogen.

Glycogen is stored as granules in the cytoplasm. Some cells are stimulated to take up glucose and use it immediately for cell respiration. These processes lower the levels of glucose in the blood.

When glucose levels are too low (below the homeostasis set point), α cells in the pancreatic islets produce glucagon. Glucagon stimulates the liver cells to convert glycogen into glucose and release it into the blood, thereby raising the blood glucose level.

Cholesterol in the body comes from the diet and *hepatic synthesis* (i.e., production by the liver). The liver excretes cholesterol by adding it to bile. The homeostatic control that keeps the plasma cholesterol level constant mainly involves hepatic synthesis.

The ingestion of saturated fatty acids (animal fats) raises plasma cholesterol levels, while the ingestion of unsaturated fatty acids (vegetable fats) lowers it. Low-density lipoproteins (LDL) deliver cholesterol to cells throughout the body.

High-density lipoproteins (HDL) remove excess cholesterol from blood and tissue and deliver it to the liver for excretion. The ratio of LDL to HDL is important, and the lower the ratio, the lower the deposition of extra cholesterol in the blood vessels.

Obesity and regulation of body mass

Diabetes mellitus, commonly called diabetes, is a disease where body cells are unable to metabolize sugar. Blood glucose levels become high enough for the kidneys to excrete glucose, which can be tested in a urine sample. The liver cannot store glucose as glycogen, and cells are unable to utilize glucose for energy. Since carbohydrates are not being metabolized, the body breaks down protein and fat for energy.

Ketones, a class of chemicals produced from the fat breakdown, then build up in the blood, resulting in reduced blood volume and acidosis (lower pH). This can lead to coma and death. In *type 1 diabetes*, usually diagnosed in children and young adults, the pancreas does not produce insulin.

A viral infection may cause cytotoxic T cells to destroy pancreatic islets. This condition of diabetics is treated with a daily administration of insulin; an overdose of insulin or lack of eating results in hypoglycemia. The brain has constant sugar requirements; low blood sugar can result in unconsciousness. An immediate intake of sugar is an effective treatment for hypoglycemia.

Of the 26 million diabetics in the U.S., the vast majority have *type 2 diabetes*. This form usually occurs in obese individuals and can occur at any age, although it generally appears later in life.

Obesity can be caused by inactivity, an unhealthy diet, lack of sleep (may cause hormonal changes), medical problems, or a combination of these. In type 2 diabetes, the pancreas does produce insulin. However, liver and muscle cells do not respond to it because they lack sufficient receptors for insulin.

Untreated, type 2 diabetes can have serious consequences, including blindness, kidney disease, circulatory disorders, and strokes. A healthy diet avoiding processed carbohydrates (low glycemic index) and regular exercise can help control the disease, and oral drugs can make cells more sensitive to insulin or stimulate the pancreas to produce higher levels of insulin, in an attempt to overcome insulin resistance.

Type 1 diabetes	Type 2 diabetes
The onset is usually early, sometime during childhood.	The onset is usually late, sometime after childhood.
β cells do not produce enough insulin.	Target cells become insensitive to insulin.
Diet by itself cannot be used to control the condition. Insulin injections are needed to control glucose levels.	Insulin injections are not mandatory. A low-fat diet can control the condition, but oral medications may be necessary.

Notes for active learning

Notes for active learning

PRACTICE QUESTIONS
&
DETAILED EXPLANATIONS

Page intentionally left blank

Practice Questions: Eukaryotic Cell: Structure & Function

1. Facilitated transport can be differentiated from active transport because:

A. active transport requires a symport

B. facilitated transport displays saturation kinetics

C. active transport displays sigmoidal kinetics

D. active transport requires an energy source

E. active transport only occurs in the mitochondrial inner membrane

2. The cell is the basic unit of function and reproduction because:

A. subcellular components cannot regenerate whole cells

B. cells can move in space

C. single cells can sometimes produce an entire organism

D. cells can transform energy to do work

E. new cells can arise by the fusion of two cells

3. Which is/are NOT able to readily diffuse through plasma membranes without a transport protein?

    I. water

    II. small hydrophobic molecules

    III. small ions

    IV. neutral gas molecules

A. I only

B. I and II only

C. I and IV only

D. III and IV only

E. I and III only

4. During a hydropathy analysis of a recently sequenced protein, a researcher discovers that the protein has several regions containing 20-25 hydrophobic amino acids. What conclusion would she draw from this finding?

A. Protein would be specifically localized in the mitochondrial inner membrane

B. Protein would be targeted to the mitochondrion

C. Protein is likely an integral protein

D. Protein is likely involved in glycolysis

E. Protein is likely secreted from the cell

5. If a membrane-bound vesicle that contains hydrolytic enzymes is isolated, it is likely a:

A. vacuole

B. microbody

C. chloroplast

D. phagosome

E. lysosome

6. A DNA damage checkpoint arrests cells in:

A. M/G2 transition

B. G1/G2 transition

C. G1/S transition

D. anaphase

E. S/G1 transition

7. If a segment of double-stranded DNA has a low ratio of guanine-cytosine (G-C) pairs relative to adenine-thymine (A-T) pairs, it is reasonable to assume that this nucleotide segment:

A. requires less energy to separate the two DNA strands than a comparable segment with a high C-G ratio

B. requires more energy to separate the two DNA strands than a comparable segment with a high C-G ratio

C. requires the same energy to separate the two DNA strands as a comparable segment with a high C-G ratio

D. contains more adenine than thymine

E. contains more cytosine than guanine

8. In general, phospholipids contain:

A. a glycerol molecule

B. saturated fatty acids

C. unsaturated fatty acids

D. a cholesterol molecule

E. a glucose molecule

9. The overall shape of a cell is determined by its:

A. cell membrane

B. cytoskeleton

C. nucleus

D. cytosol

E. endoplasmic reticulum

10. Which molecule generates the greatest osmotic pressure when placed into water?

A. 300 mM NaCl

B. 250 mM CaCl$_2$

C. 500 mM glucose

D. 600 mM urea

E. 100 mM KCl

11. Which cellular substituent is produced within the nucleus?

    A. Golgi apparatus   **C.** ribosome

    B. lysosome   **D.** rough endoplasmic reticulum

                                     E. cell membrane

12. In the initial stages of the cell cycle, progression from one phase to the next is controlled by:

    A. p53 transcription factors   **C.** origin recognition complexes

    B. anaphase-promoting complexes   **D.** pre-replication complexes

                                     E. cyclin–CDK complexes

13. Which is the correct sequence occurring during polypeptide synthesis?

    A. DNA generates tRNA → tRNA anticodon binds to the mRNA codon in the cytoplasm → → tRNA is carried by mRNA to ribosomes, causing amino acids to join in a specific order

    B. DNA generates mRNA → mRNA moves to the ribosome → tRNA anticodon binds to the mRNA codon, causing amino acids to join in their appropriate order

    C. Specific RNA codons cause amino acids to line up in a specific order → tRNA anticodon attaches to mRNA codon → rRNA codon causes the protein to cleave into specific amino acids

    D. DNA regenerates mRNA in the nucleus → mRNA moves to the cytoplasm and attaches to the tRNA anticodon → operon regulates the sequence of amino acids in the appropriate order

    E. DNA generates mRNA → mRNA anticodon binds to tRNA codon causing amino acids to join together in their appropriate order

14. Both prokaryotes and eukaryotes contain:

    I. a plasma membrane   II. ribosomes   III. peroxisomes

    A. I only   **C.** I and II only

    B. II only   **D.** II and III only

                                     E. I, II and III

15. RNA is NOT expected to be in a:

    A. nucleus   **C.** prokaryotic cell

    B. mitochondrion   **D.** ribosome

                                     E. vacuole

16. Phosphotransferase is needed to form the mannose-6-phosphate tag that targets hydrolase enzymes to their lysosomal destination. Defective phosphotransferase causes I-cell disease, whereby the defective organelle which gives rise to this condition is the:

A. nucleus

B. cell membrane

C. Golgi apparatus

D. smooth ER

E. nucleolus

17. Mitochondria and chloroplasts are unusual organelles because they:

A. synthesize all their ATP using substrate-level phosphorylation

B. contain cytochrome C oxidase

C. are devoid of heme-containing proteins

D. contain nuclear-encoded and organelle-encoded proteins

E. degrade macromolecules using hydrolytic enzymes

18. All statements are true about cytoskeleton EXCEPT that it:

A. is not required for mitosis

B. maintains the cell's shape

C. gives the cell mechanical support

D. is composed of microtubules and microfilaments

E. is important for cell motility

19. Inside the cell, several key events in the cell cycle include:

I. DNA damage repair and replication completion

II. centrosome duplication

III. assembly of the spindle and attachment of the kinetochores to the spindle

A. I only

B. I, II and III

C. I and II only

D. II and III only

E. I and III only

20. A researcher labeled *Neurospora* mitochondria with a radioactive phosphatidylcholine membrane component and followed cell division by autoradiography in an unlabeled medium, allowing enough time for one cell division. What results did this scientist observe before concluding that pre-existing mitochondria give rise to new mitochondria?

    A. Daughter mitochondria are labeled equally

    B. Daughter mitochondria are all unlabeled

    C. One-fourth of daughter mitochondria are labeled

    D. Some of the daughter mitochondria are unlabeled, while some are labeled

    E. Daughter mitochondria are all labeled

21. Which involves the post-translational modification of proteins?

    A. peroxisomes   **C.** Golgi complex

    B. vacuoles   **D.** lysosomes

                                    E. smooth ER

22. The width of a typical animal cell is closest to:

    A. 1 millimeter   **C.** 1 micrometer

    B. 20 micrometers   **D.** 10 nanometers

                                      E. 100 micrometers

23. Which organelle is identified by the sedimentation coefficient – S units (Svedberg units)?

    A. peroxisome   **C.** mitochondrion

    B. nucleus   **D.** nucleolus

                                      E. ribosome

24. Which is NOT involved in osmosis?

    A. H_2O spontaneously moves from a hypertonic to a hypotonic environment

    B. H_2O spontaneously moves from an area of high solvent to a low solvent concentration

    C. H_2O spontaneously moves from a hypotonic to a hypertonic environment

    D. Transport of H_2O

    E. Diffusion of H_2O

25. During cell division, cyclin B is marked for destruction by the:

A. p53 transcription factor

B. anaphase-promoting complex

C. CDK complex

D. pre-replication complex

E. maturation promoting factor

26. When a female mouse with a defect in a mitochondrial protein required for fatty acid oxidation is crossed with a wild-type male, all progeny (both male and female) have the wild-type phenotype. Which statement is likely correct?

A. Mice do not exhibit maternal inheritance

B. The defect is a result of an autosomal X-linked recessive trait

C. The defect is a result of an X-linked recessive trait

D. The defect is a result of a recessive mitochondrial gene

E. The defect is a result of a nuclear gene mutation

27. The smooth ER participates in:

A. substrate-level phosphorylation

B. exocytosis

C. synthesis of phosphatidylcholine

D. allosteric activation of enzymes

E. synthesis of cytosolic proteins

28. What type of organelle is in plants but not in animals?

A. ribosomes

B. mitochondria

C. nucleus

D. plastids

E. nucleolus

29. Mitochondrial mutations are often limited to one tissue type. If an individual does not produce blood calcium-decreasing hormone calcitonin, which tissue is likely to carry the mitochondrial mutation?

A. thyroid

B. kidney

C. parathyroid

D. liver

E. spleen

30. The rough ER participates in:

A. oxidative phosphorylation

B. synthesis of plasma membrane proteins

C. endocytosis

D. post-translational modification of enzymes

E. synthesis of lipids

31. The best definition of active transport is the movement of:

A. solutes across a semipermeable membrane down an electrochemical gradient

B. solutes across a semipermeable membrane up a concentration gradient

C. substances across a membrane per the Donnan equilibrium

D. solutes via osmosis across a semipermeable membrane from high to low concentration

E. H_2O via diffusion across a semipermeable membrane

32. Overexpression of cyclin D:

A. increases contact inhibition

B. decreases telomerase activity

C. activates apoptosis

D. promotes unscheduled entry into the S phase

E. promotes the transition from G1 to the S phase

33. A failure in which stage of spermatogenesis produces nondisjunction and males with XXY karyotype?

A. prophase I

B. metaphase

C. prophase II

D. telophase

E. anaphase I

34. Protein targeting occurs during the synthesis of which type of proteins?

A. nuclear proteins

B. secreted proteins

C. cytosolic proteins

D. mitochondrial proteins

E. chloroplast proteins

35. The difference between "free" and "attached" ribosomes is that:

    I. Free ribosomes are in the cytoplasm, while attached ribosomes are anchored to the endoplasmic reticulum

    II. Free ribosomes produce proteins in the cytosol, while attached ribosomes produce proteins that are inserted into the ER lumen

    III. Free ribosomes produce proteins that are exported from the cell, while attached ribosomes make proteins for mitochondria and chloroplasts

A. I only

B. II only

C. I and II only

D. I, II and III

E. I and III only

36. The concentration of growth hormone receptors will significantly reduce after selective destruction of which structure?

A. nucleolus

B. nucleus

C. cytosol

D. plasma membrane

E. peroxisomes

37. All these processes are ATP-dependent, EXCEPT:

A. export of Na^+ from a neuron

B. influx of Ca^{2+} into a muscle cell

C. influx of K^+ into a neuron

D. exocytosis of neurotransmitter at a nerve terminus

E. movement of urea across a cell membrane

38. The loss of function of p53 protein results in:

A. blockage in activation of the anaphase-promoting complex

B. increase of contact inhibition

C. elimination of the DNA damage checkpoint

D. activation of apoptosis

E. suppression of spindle fiber assembly

39. Which organelle is most closely associated with exocytosis of newly synthesized secretory protein?

A. peroxisome

B. ribosome

C. lysosome

D. Golgi apparatus

E. nucleus

40. Plant membranes are more fluid than animal membranes because plant membranes:

A. contain substantial amounts of cholesterol

B. have higher amounts of unsaturated fatty acids compared to the membranes of animals

C. have lower amounts of unsaturated fatty acids compared to the membranes of animals

D. are only found in the inner membrane of the mitochondria

E. do not contain glycosylated proteins

41. Which organelles are enclosed in a double membrane?

I. nucleus II. chloroplast III. mitochondrion

A. I and II only

B. II and III only

C. I and III only

D. I, II and III

E. I only

42. Proteins are marked and delivered to specific cell locations through:

A. specific protein transport channels

B. regulation signals released by the cell's cytoskeleton

C. post-translational modifications occurring in the Golgi

D. compartmentalization of the rough ER during protein synthesis

E. post-translational modifications occurring in the nucleus

43. The presence of which element differentiates a protein from a carbohydrate molecule?

A. carbon

B. hydrogen

C. nitrogen

D. oxygen

E. nitrogen and oxygen

44. What change occurs in the capillaries when arterial blood is infused with the plasma protein albumin?

    A. Decreased movement of H_2O from the capillaries into the interstitial fluid
    B. Increased movement of H_2O from the capillaries into the interstitial fluid
    C. Decreased movement of H_2O from the interstitial fluid into the capillaries
    D. Increased permeability to albumin
    E. Decreased permeability to albumin

45. The retinoblastoma protein controls:

    A. contact inhibition
    B. transition from G1 to S phase
    C. activation of apoptosis
    D. expression of cyclin D
    E. A and C

46. All processes take place in the mitochondrion, EXCEPT:

    A. oxidation of pyruvate
    B. Krebs cycle
    C. electron transport chain
    D. reduction of FADH into $FADH_2$
    E. glycolysis

47. Which is the most abundant lipid in the human body?

    A. teichoic acid
    B. peptidoglycan
    C. glycogen
    D. triglycerides
    E. glucose

48. What is the secretory sequence in the flow of newly synthesized protein for export from the cell?

    A. Golgi → rough ER → smooth ER → plasma membrane
    B. Golgi → rough ER → plasma membrane
    C. smooth ER → rough ER → Golgi → plasma membrane
    D. rough ER → smooth ER → Golgi → plasma membrane
    E. rough ER → Golgi → plasma membrane

49. Digestive lysosomal hydrolysis would affect all the following EXCEPT:

    A. proteins
    B. minerals
    C. nucleotides
    D. lipids
    E. carbohydrates

50. Recycling of organelles within the cell is accomplished through autophagy by:

A. mitochondria

B. peroxisomes

C. nucleolus

D. lysosomes

E. rough endoplasmic reticulum

51. All are lipid derivatives EXCEPT:

A. carotenoids

B. albumins

C. waxes

D. steroids

E. lecithin

52. Defective attachment of a chromosome to the spindle:

A. blocks activation of the anaphase-promoting complex

B. activates exit from mitosis

C. prevents overexpression of cyclin D

D. activates sister chromatid separation

E. promotes the transition from G1 to the S phase

53. Which stage is when human cells with a single unreplicated copy of the genome are formed?

A. mitosis

B. meiosis II

C. meiosis I

D. interphase

E. G1

54. Which organelle in the cell is the site of fatty acid, phospholipid, and steroid synthesis?

A. endosome

B. peroxisome

C. rough endoplasmic reticulum

D. smooth endoplasmic reticulum

E. chloroplast

55. Which eukaryotic organelle is NOT membrane-bound?

A. nucleus

B. plastid

C. centriole

D. chloroplast

E. C and D

56. Which phase of mitotic division do spindle fibers split the centromere and separate the sister chromatids?

A. interphase

B. telophase

C. prophase

D. metaphase

E. anaphase

57. All are correct about cyclic AMP (cAMP), EXCEPT:

A. the enzyme that catalyzes the formation of cAMP is in the cytoplasm

B. membrane receptors can activate the enzyme that forms cAMP

C. ATP is the precursor molecule in the formation of cAMP

D. adenylate cyclase is the enzyme that catalyzes the formation of cAMP

E. cAMP is a second messenger that triggers a cascade of intracellular reactions when a peptide hormone binds to a receptor on the cell membrane

58. Which molecule, and its associated protein kinase, ensures a proper progression of cell division?

A. oncogene

B. tumor suppressor

C. cyclin

D. histone

E. homeotic

59. Placed in a hypertonic solution, erythrocytes will undergo:

A. crenation

B. expansion

C. plasmolysis and rupture

D. no change

E. shrinkage and then rapid expansion

60. Tiny organelles abundant in the liver contain oxidases and detoxify substances (i.e., alcohol and hydrogen peroxide) are:

A. centrosome

B. lysosomes

C. endosomes

D. rough endoplasmic reticulum

E. peroxisomes

61. Microtubules can function independently or form protein complexes to produce structures like:

    I. flagella

    II. actin and myosin filaments in muscle cells

    III. mitotic spindle apparatus

A. III only

B. I and III only

C. II and III only

D. I and II only

E. I, II, and III

62. About a cell that secretes much protein (e.g., a pancreatic exocrine cell), it can be assumed that this cell has:

    I. an abundance of rough endoplasmic reticulum

    II. a large Golgi apparatus

    III. a prominent nucleolus

A. I only

B. I and III only

C. II and III only

D. I, II and III

E. I and II only

63. A cell division where each daughter cell receives a chromosome complement identical to the parent is:

A. mitosis

B. non-disjunction

C. meiosis

D. replication

E. crossing-over

64. Which is affected LEAST by colchicine, which interferes with microtubule formation?

A. organelle movement

B. meiosis

C. mitosis

D. cilia

E. pseudopodia for amoeboid motility

65. Which process is NOT an example of apoptosis?

A. Reabsorption of a tadpole's tail during metamorphosis into a frog

B. Formation of the endometrial lining of the uterus during the menstrual cycle

C. Formation of the synaptic cleft by triggering cell death in brain neuronal cells

D. Formation of fingers in the fetus by the removal of tissue between the digits

E. All are examples of apoptosis

66. Cells that utilize large quantities of ATP for active transport (e.g., epithelial cells of the intestine) have:

A. many mitochondria

B. elevated levels of DNA synthesis

C. high levels of adenylate cyclase

D. polyribosomes

E. many lysosomes

67. Clathrin is a protein collected on the cytoplasmic side of cell membranes and functions in the coordinated pinching off membrane into receptor-mediated endocytosis. It can be predicted that a lipid-soluble toxin that inactivates clathrin results in:

A. increased ATP consumption

B. increased protein production on the rough endoplasmic reticulum

C. increased secretion of hormone into the extracellular fluid

D. increased ATP production

E. reduced delivery of polypeptide hormones to endosomes

68. Cells that respond to peptide hormones usually do so through biochemical reactions involving membrane receptors and kinase activation. For cells to respond, the first and second messengers communicate because:

A. peptide hormones pass through the cell membrane and elicit a response

B. the hormone-receptor complex moves into the cytoplasm as a unit

C. hormones alter cellular activities directly through gene expression

D. the G protein acts as a link between the first and second messengers

E. the presence of a cytosolic receptor binds the hormone

69. Mitosis does NOT serve the purpose of:

A. replenishment of erythrocytes

B. formation of scar tissue

C. organ repair

D. tissue growth

E. transduction

Notes or active learning

Notes or active learning

Detailed Explanations: Eukaryotic Cell: Structure & Function

Answer Key

1: D	11: C	21: C	31: B	41: D	51: B	61: B
2: A	12: E	22: B	32: D	42: C	52: A	62: D
3: E	13: B	23: E	33: E	43: C	53: B	63: A
4: C	14: C	24: A	34: B	44: A	54: D	64: E
5: E	15: E	25: B	35: C	45: B	55: C	65: B
6: C	16: C	26: E	36: D	46: E	56: E	66: A
7: A	17: D	27: C	37: E	47: D	57: A	67: E
8: A	18: A	28: D	38: C	48: E	58: C	68: D
9: B	19: B	29: A	39: D	49: B	59: A	69: E
10: B	20: A	30: B	40: B	50: D	60: E	

1. D is correct.

Facilitated transport (or *passive transport*) uses a protein pore (or channel) but is differentiated from *active transport, which requires energy.*

2. A is correct.

Cell is the unit of function and reproduction because subcellular components cannot regenerate whole cells.

3. E is correct.

I: water relies on *aquaporins* to readily diffuse across a plasma membrane.

Without aquaporins, only a small fraction of water molecules can diffuse through the cell membrane per unit of time because of the polarity of water molecules.

II: *small hydrophobic molecules* readily diffuse through the hydrophobic tails of the plasma membrane.

III: *small ions* rely on ion channel transport proteins to diffuse across the membrane.

IV: *neutral gas molecules* (e.g., O_2, CO_2) readily diffuse through the hydrophobic tails of the plasma membrane.

4. C is correct.

Integral proteins often span the plasma membrane.

Plasma membrane is a phospholipid bilayer with an inner span of hydrophobic (or *water-fearing*) regions.

Diameter of a plasma membrane is 20 to 25 amino acids thick.

Spans of 20-25 hydrophobic residues prefer to be embedded in the plasma membrane and isolated from water.

Hydropathy analysis determines the degree of *hydrophobicity* (i.e., nonpolar or *water-fearing*) or *hydrophilicity*
(i.e., polar or *water-loving*) of amino acids in a protein. It characterizes the possible structures of a protein.

Each residue (i.e., individual amino acid) has a hydrophobicity value (analogous to electronegativity).

Hydropathy analysis plots the degree of hydrophobicity (or *hydrophilicity*) on the *y*-axis and the amino acid sequence on the *x*-axis, *hydrophobicity vs. amino acid* position.

Amino acids have lower energy when polar amino acids occupy a polar environment (e.g., cytoplasm) and nonpolar amino acids occupy a nonpolar environment (e.g., the interior of the plasma membrane).

5. E is correct.

Lysosome is the digestive region of the cell and is a membrane-bound organelle with a low pH (around 5) that stores hydrolytic enzymes.

A: *vacuoles* and *vesicles* are membrane-bound sacs involved in the transport and storage of materials ingested, secreted, processed, or digested by cells.

Vacuoles are larger than vesicles and are in plant cells (e.g., central vacuole).

C: *chloroplasts* are the site of photosynthesis and are only in algae and plant cells.

Like mitochondria, chloroplasts contain circular DNA and ribosomes and may have evolved similarly via endosymbiosis.

D: *phagosomes* are vesicles for transporting and storing materials ingested by the cell through phagocytosis. Vesicles form by the fusion of the cell membrane around the particle.

Phagosome is a cellular compartment in which pathogenic microorganisms are digested.

Phagosomes *fuse with lysosomes* in their maturation process to form *phagolysosomes*.

6. C is correct.

DNA damage checkpoints are *signal transduction pathways* that block cell cycle progression in G1, G2, and metaphase and slow the S phase progression rate when DNA is damaged.

Pausing cell cycle allows the cell to repair the damage before dividing.

$$G1 \rightarrow S \rightarrow G2 \rightarrow prophase \rightarrow metaphase \rightarrow anaphase \rightarrow telophase$$
$$\underbrace{\qquad\qquad}_{Interphase} \qquad \underbrace{\qquad\qquad\qquad\qquad\qquad}_{Mitosis\,(PMAT)}$$

Cell cycle divides into interphase (G1. S, G2) and mitosis (prophase, metaphase, anaphase, and telophase)

7. A is correct.

(C≡G) base pairs are linked in the double helix by *three hydrogen bonds*.

(A=T) base pairs are linked in the double helix by *two hydrogen bonds*.

Therefore, it takes more energy to separate G-C base pairs.

Less G–C rich strands of double-stranded DNA require less energy to separate (i.e., denature).

Chargaff's rule specifies complementary base pairing; double-stranded DNA has equal G and C (and A and T).

8. A is correct.

Phospholipids are lipids as major components of cell membranes; they form lipid bilayers.

Phospholipids contain a glycerol backbone, a phosphate group, and a simple organic molecule (e.g., choline).

The 'head' is *hydrophilic* (i.e., attracted to water), while the 'tails' are *hydrophobic* (i.e., repelled by water), and the tails aggregate (via hydrophobic forces).

Hydrophilic heads contain negatively charged phosphate groups and glycerol.

Hydrophobic tails usually consist of 2 long fatty acids (saturated or unsaturated) hydrocarbon chains.

Cholesterol is embedded within animal lipid bilayers but is absent in plant cell membranes. Embedded cholesterol in the phospholipid bilayer in eukaryotic animal cells allows for protective cell membranes that can also change shape.

Plant and bacterial cell walls are primarily composed of cellulose and peptidoglycan glucose polymers, respectively, which are rigid and restrict movement.

9. B is correct.

Cytoskeleton determines the *overall shape* of a cell and is composed of:

> *Microtubules* help synthesize cell walls in plants which primarily contribute to shape; centrosomes organize microtubules, cilia, and flagella (9+2).

> Microtubules form *mitotic spindles* and *centrioles* (non-membrane-bound organelles).

> *Intermediate filaments*: anchor organelles (e.g., nucleus) and bear tension, contributing to cell shape.

> *Microfilaments:* resist compression, thus contributing to cell shape.

10. B is correct.

Osmolarity is determined by the total concentration of dissolved particles in the solution.

Compounds that dissociate into ions (i.e., electrolytes) increase the concentration of particles and produce a higher osmolarity.

To determine which molecule (after dissociation into ions) generates the highest osmolarity, determine the number of individual ions each molecule dissociates into in H_2O.

> $CaCl_2$ dissociates into 1 Ca^{2+} and 2 Cl^-

As a result, $CaCl_2$ generates the greatest osmolarity, which equals 250 mOsmoles for Ca^{2+} + 500 mOsmoles for $Cl^- = 750$ mOsmoles when $CaCl_2$ dissociates into ions within the solution.

A: NaCl dissociates in water with an osmolarity of 600 mOsmoles because it dissociates into 1 Na^+ and 1 Cl^-.

Na^+ cations and Cl^- anions are added: 300 mOsmoles for Na^+ + 300 mOsmoles for $Cl^- = 600$ mOsmoles.

C: glucose does *not* dissociate in water and has the same osmolarity as the starting molecule—500 mOsmoles.

D: urea does *not* dissociate in water and has the same osmolarity as the starting molecule—600 mOsmoles.

11. C is correct.

The two ribosomal subunits are synthesized in the nucleolus, a region within the nucleus. The ribosomes are the sites of protein production.

Prokaryotic ribosomes (30S small + 50S large subunit = 70S complete ribosome) are smaller than *eukaryotic ribosomes* (40S small + 60S large subunit = 80S complete ribosome).

A: *Golgi apparatus* is a membrane-bound organelle that modifies (e.g., glycosylation), sorts, and packages proteins synthesized by the ribosomes.

B: *lysosomes* have a low pH of about 5 and contain hydrolytic enzymes involved in digestion.

D: *rough endoplasmic reticulum* (RER) is part of the endomembrane extending from the nuclear envelope.

RER has ribosomes associated with its membrane and is the site of the production and folding of proteins.

Misfolded proteins exit the rough ER and are sent to the *proteasome* for degradation.

E: *cell membrane* is a barrier between the interior and exterior of the cell.

12. E is correct.

Cyclins are phosphorylated proteins responsible for specific events during cycle division, such as microtubule formation and chromatin remodeling.

Cyclins are four classes based on their behavior in the cell cycle: G1/S, S, M, and G1 cyclins.

p53 is not a transcription factor but is a tumor suppressor gene. The p53 protein is crucial in multicellular organisms, where it regulates the cell cycle and functions as a tumor suppressor (preventing cancer).

p53 is *the guardian of the genome* because it conserves stability by preventing genome mutation.

13. B is correct.

Codon is a three-nucleotide segment of an mRNA that hybridizes (via complementary base pairing) with the appropriate anticodon on the tRNA to encode one amino acid in a polypeptide chain during protein synthesis.

The tRNA molecule interacts with the mRNA codon after the ribosomal complex binds the mRNA.

A: tRNA molecules interact with the mRNA codon after (not before) the ribosomal complex binds mRNA

C: *translation* involves the conversion of mRNA into protein.

D: *operons* regulate the transcription of genes into mRNA and are not involved in translating mRNA into proteins.

14. C is correct.

Peroxisomes are organelles in most eukaryotic cells; a significant function is the breakdown of very-long-chain fatty acids through *beta-oxidation*.

Peroxisomes convert the very long fatty acids to medium-chain fatty acids in animal cells, subsequently shuttled to the mitochondria. Medium-chain fatty acids are degraded, via oxidation, into CO_2 and H_2O.

Peroxisomes are membrane-bound, while ribosomes are non-membrane-bound organelles.

15. E is correct.

Vacuole is a membrane-bound organelle in plant and fungal cells and some protist, animal, and bacterial cells.

Vacuoles are enclosed compartments filled with water. They contain inorganic and organic molecules (including enzymes in solution) and may contain solids that have been engulfed.

Function and significance of vacuoles vary by cell type, with much greater prominence in plants, fungi, and certain protists than in animals or bacteria.

Mitochondria and chloroplasts have circular DNA resembling DNA in prokaryotes, as supported by the endosymbiotic *theory* for the evolution of mitochondria and chloroplasts from prokaryotes.

Ribosomes are the site of protein synthesis.

16. C is correct.

I-cell disease patients cannot correctly direct newly synthesized peptides to their target organelles. The Golgi apparatus is part of the endomembrane system and serves as a cellular distribution center.

Golgi apparatus packages proteins before they are sent to their destination. It modifies, sorts, and packages proteins for cell secretion (i.e., exocytosis) or use within the cell.

A: *nucleus* is the largest organelle, contains genetic material (i.e., DNA), and is the site of rRNA synthesis (within the nucleolus).

D: *smooth ER* (endoplasmic reticulum) forms the endomembrane system, is connected to the nuclear envelope, and functions in several metabolic processes. It synthesizes lipids, phospholipids, and steroids.

Cells secreting lipids, phospholipids, and steroids (e.g., testes, ovaries, and skin oil glands) have an extensive smooth endoplasmic reticulum.

Smooth ER conducts the metabolism of carbohydrates and drug detoxification.

It is responsible for the attachment of receptors on cell membrane proteins and steroid metabolism.

E: *nucleolus* is the organelle within the nucleus responsible for synthesizing ribosomal RNA (rRNA).

17. D is correct.

Cytochrome c oxidase is a large transmembrane protein complex in bacteria and the mitochondrion of eukaryotes. It receives an electron from each of four cytochrome c molecules and transfers the electrons to an O_2 molecule, converting molecular oxygen to two molecules of H_2O.

Cytochrome c oxidase is the *last enzyme in the electron transport chain* (ETC) of mitochondria (or bacteria) in the mitochondrial (or bacterial) inner membrane.

Mitochondria and chloroplasts have circular DNA resembling DNA in prokaryotes.

Endosymbiotic theory illustrates the evolution of mitochondria and chloroplasts from prokaryotes, thus explaining the resemblance.

18. A is correct.

Cytoskeleton is integral to proper cell division because it forms the mitotic spindle and separates sister chromatids during cell division.

Cytoskeleton comprises microtubules and microfilaments, provides mechanical cell support to maintain shape, and functions in cell motility.

19. B is correct.

Biochemical events during the cell cycle include DNA damage repair and replication completion, centrosome duplication, spindle assembly, and attachment of the kinetochores to the spindle.

20. A is correct.

Mitochondria divide autonomously (i.e., independent of the genome) to produce daughter mitochondria that incorporate new nonradioactive phosphatidylcholine and inherit radioactive phosphatidylcholine from the parent via *semiconservative replication*.

Therefore, the daughter mitochondria have equal radioactivity.

B: mitochondria divide *autonomously*, and daughter mitochondria retain parental (original) radioactive label.

C: original sample was 100% radiolabeled.

DNA replication is *semiconservative*, whereby one strand of parental DNA (radiolabeled) and one strand of newly replicated (non-radiolabeled) are in each daughter cell after the first round of division.

D: requires the daughter mitochondria to be synthesized *de novo* (i.e., new) with newly synthesized, nonradioactive phosphatidylcholine.

If the mitochondria divide autonomously, the daughter mitochondria retain the radioactive label evenly via *semiconservative replication*.

21. C is correct.

Golgi apparatus (i.e., Golgi complex) is a eukaryotic cell organelle.

Golgi apparatus processes proteins via *post-translational modifications* for three destinations:

 1) *secreted* from the cell,

 2) transported into *organelles,* or

 3) targeted to the *plasma membrane.*

Golgi complex mainly processes proteins synthesized by the ER

In plant cells, the central vacuole functions as a lysosome, stores nutrients and maintains osmotic balance.

Peroxisomes are like lysosomes in size, are bound by a single membrane, and are filled with enzymes.

However, peroxisomes *bud from* the endoplasmic reticulum.

22. B is correct.

Prokaryotic cells (bacteria) have a typical cell width of 0.2 to 2.0 micrometers in diameter.

Eukaryotic cells (animal cells) have a typical cell width of 10-100 micrometers in diameter.

 1 millimeter = 1×10^{-3} m

 1 micrometer = 1×10^{-6} m

 1 nanometer = 1×10^{-9} m

Light microscopes visualize objects from 1 millimeter (10^{-3} m) to 0.2 micrometers (2×10^{-7} m).

Electron microscopes visualize objects as small as an atom (1 angstrom or 10^{-10} m).

Microscopic scale ranges from 1 millimeter (10^{-3} m) to a ten-millionth of a millimeter (10^{-10} m). There are immense variations in objects' sizes, even within the microscopic scale.

10^{-3} m is 10 million times larger than 10^{-10} m, equivalent to the Earth's size *vs.* a beach ball.

                    Copyright © Sterling Education

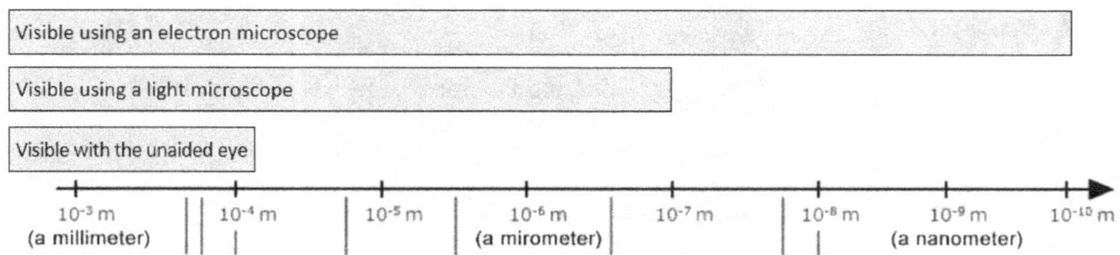

Comparison of resolution by unaided eyes, light, and electron microscopes

23. E is correct.

Ribosomes are composed of specific rRNA molecules and associated proteins.

Ribosomes are identified by the sedimentation coefficients (i.e., S units for Svedberg units) for density.

Prokaryotes have a 30S small and a 50S large subunit (i.e., complete ribosome = 70S; based on density).

Eukaryotes have a 40S small and 60S large subunit (i.e., complete ribosome = 80S).

A: *peroxisomes* are organelles involved in hydrogen peroxide (H_2O_2) synthesis and degradation. They function in cell detoxification and contain the catalase that decomposes H_2O_2 into H_2O and O_2.

C: *mitochondria* are organelles as the site of cellular respiration (i.e., oxidation of glucose to yield ATP) and plentiful in cells with high demands for ATP (e.g., muscle cells).

The number of mitochondria within a cell varies widely by organism and tissue type.

Many cells have a single mitochondrion, whereas others contain several thousand mitochondria.

Nucleus is the largest membrane-bound organelle in eukaryotes, containing the genetic code (i.e., DNA).

It directs the storing and transmitting of genetic information. Cells can contain multiple nuclei (e.g., skeletal muscle cells), one nucleus, or none (e.g., red blood cells).

24. A is correct.

Osmosis is a type of diffusion involving water and is a form of passive transport.

Hypertonic means a solution of high solute and low solvent concentrations.

Hypotonic means a solution of high solvent and low solute concentrations.

Solvents flow spontaneously from an area of high solvent to a low solvent concentration.

During *osmosis*, water flows from a hypotonic to a hypertonic environment.

25. B is correct.

Complex of Cdk and cyclin B is a maturation or mitosis-promoting factor (MPF).

Cyclin B is necessary to progress cells into and out of the M phase of the cell cycle.

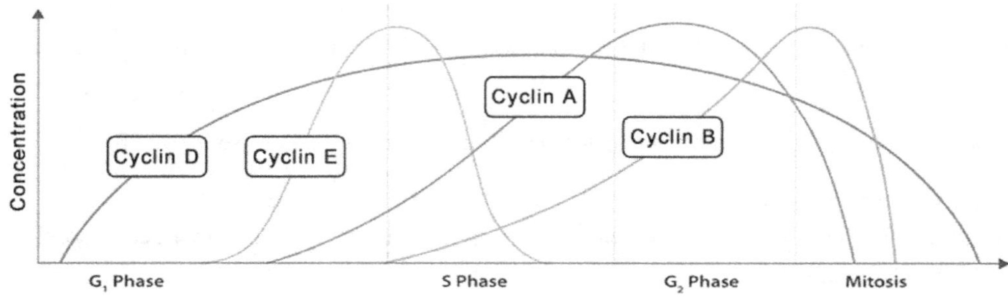

26. E is correct.

Mitochondria have their DNA genetic material and machinery to manufacture their RNAs and proteins.

For example, the trait is recessive (not observed in limited number of offspring) and encoded by a nuclear gene.

A: like organisms that reproduce sexually, mice inherit the *mitochondrial organelle* from their mother and display maternal inheritance of mitochondrial genes.

C: if X-linked, selectively male (not female) progeny would display the trait.

D: mitochondrial genes cannot be *recessive* because mitochondria are *inherited from the mother*.

27. C is correct.

Smooth endoplasmic reticulum (smooth ER) participates in synthesizing phosphatidylcholine.

Smooth ER functions include:

1) synthesis of lipids, phospholipids, and steroids

2) metabolism of carbohydrates and steroids

3) detoxifying alcohol and drugs

4) regulating Ca^{2+} concentration in muscle cells

Phosphatidylcholine is a class of phospholipids.

28. D is correct.

Plastids (e.g., chloroplast and chromoplast) are major organelles in plants and algae.

Plastids are the site of manufacturing and storing critical chemical compounds used by cells.

They often contain pigments used in photosynthesis. Pigments change to determine cell color.

Plastids, like prokaryotes, contain a circular double-stranded DNA molecule.

29. A is correct.

Thyroid gland synthesizes calcitonin in response to high blood calcium levels. It acts to reduce blood calcium (Ca^{2+}), opposing the effects of parathyroid hormone (PTH).

Calcitonin lowers blood Ca^{2+} levels in three ways:

1) inhibiting Ca^{2+} absorption by the intestines,

2) inhibiting osteoclast activity in bones, and

3) inhibiting renal tubular cell reabsorption of Ca^{2+}, allowing it to be excreted in the urine.

B: *kidneys* serve several essential regulatory roles. They are essential in the urinary system and serve homeostatic functions such as regulating electrolytes, maintaining acid-base balance, and regulating blood pressure (maintaining salt and water balance).

Kidneys secrete *renin* (involved in blood pressure regulation) that induces the release of aldosterone from the adrenal cortex (it increases blood pressure via sodium reabsorption, increasing blood pressure).

C: *parathyroid glands* synthesize parathyroid hormone (PTH), which increases blood calcium.

PTH increases calcium concentration in blood by acting upon the parathyroid hormone receptor (elevated levels in bone and kidney) and the parathyroid hormone receptor (elevated levels in the central nervous system, pancreas, testes, and placenta).

D: *liver* is the largest organ with many functions, including detoxification, protein synthesis, and the production of biomolecules necessary for digestion. The liver synthesizes bile, which is necessary for dietary lipid emulsification (in the small intestine).

E: *spleen* is a *reservoir* for red blood cells and *filters* the blood.

30. B is correct.

Rough endoplasmic reticulum (rough ER) participates in synthesizing plasma membrane proteins.

Oxidative phosphorylation is a series of redox reactions in the electron transport chain (ETC), leading to the production of ATP. In Eukaryotes, this occurs in the *mitochondrial inner membrane* and in prokaryotes in the intermembrane space.

Endocytosis is when a cell membrane invaginates, forming a vacuole to store molecules from the extracellular space actively transported across the cell's plasma membrane.

Post-translational modification is after ribosomes translate mRNA into polypeptide chains; the polypeptide chains become a mature protein by undergoing biochemical reactions (e.g., cleavage, folding).

31. B is correct.

Active transport uses a carrier protein and energy to move a substance across a membrane against (i.e., up) a concentration gradient: from low solute to a region of high solute concentration.

Donnan equilibrium refers to some ionic species passing through the barrier while others cannot.

Charged substances that cannot pass through the membrane create an uneven electrical charge.

Donnan potential is the electric potential arising between the solutions.

32. D is correct.

Cyclin D synthesis is initiated during G1 and drives the G1/S phase transition.

Apoptosis is programmed cell death (PCD) that may occur in multicellular organisms. Biochemical events lead to characteristic cell changes (morphology) and death.

Apoptotic changes include blebbing (i.e., an irregular bulge in the plasma membrane), cell shrinkage, chromatin condensation, nuclear fragmentation, and chromosomal DNA fragmentation.

In contrast to necrosis, traumatic cell death resulting from acute cellular injury, apoptosis confers advantages during an organism's life cycle.

For example, the differentiation of fingers and toes in a developing human embryo occurs because cells between the fingers undergo apoptosis, and the digits are separated.

Unlike necrosis, apoptosis produces cell fragments called *apoptotic bodies* that phagocytic cells can engulf and quickly remove before the cell's contents can spill out onto surrounding cells and cause damage.

33. E is correct.

During meiosis I, *homologous chromosomes* separate.

Sister chromatids (identical copies, except for recombination) separate during meiosis II.

Klinefelter syndrome (XXY karyotype) contains two X and one Y chromosome.

X and Y would be "homologous chromosomes" and typically separate during meiosis I.

Failure to separate during meiosis I could create a sperm containing an X and a Y, causing Klinefelter syndrome.

Anaphase is when the *centromere splits*, and the *homologous chromosomes / sister chromatids* are drawn away (via spindle fibers) from each other toward opposite sides of the two cells.

Homologous chromosomes separate during anaphase I, while sister chromatids separate during anaphase II.

In females, Turner's syndrome is due to the single X karyotype (single X chromosome and lacking a Y).

34. B is correct.

Most proteins that are secretory, membrane-bound, or targeted to an organelle use the *N-terminal signal sequence* (i.e., 5 to 30 amino acids) to target the protein.

Signal sequence of the polypeptide is recognized by a signal recognition particle (SRP), while the protein is synthesized on ribosomes.

Synthesis pauses while the ribosome-protein complex transfers to an SRP receptor on the ER (in eukaryotes) or plasma membranes (in prokaryotes) before polypeptide translation resumes.

35. C is correct.

Ribosomes anchored to the endoplasmic reticulum create the *rough endoplasmic reticulum* "studded" with ribosomes in contrast to the *smooth endoplasmic reticulum*, which lacks ribosomes.

"*Free*" and "*attached*" ribosome comparisons:

> 1) *free ribosomes* are in the cytoplasm, while *attached ribosomes* are anchored to the endoplasmic reticulum (ER), and

> 2) *free ribosomes* produce proteins in the cytosol, while *attached ribosomes* produce proteins inserted into the ER lumen (i.e., interior space).

36. D is correct.

Anterior pituitary hormones (including GH) are peptides. Peptide hormones are *hydrophilic* and *cannot* cross the hydrophobic phospholipid bilayer; peptide hormones bind to receptors on the plasma membrane.

Destruction of the plasma membrane dramatically *reduces* the concentration of GH receptors.

37. E is correct.

Urea, a byproduct of amino acid metabolism, is a small uncharged molecule that crosses cell membranes by simple diffusion – a passive process that does not require energy.

A and C: export of Na^+ from a neuron is coupled with the *import of K^+*; the sodium-potassium-ATPase pump is an ATP-dependent process necessary to maintain a voltage potential across the neuron membrane.

B: movement of Ca^{2+} into a muscle cell occurs against the concentration gradient, and Ca^{2+} enters cells by *active transport*, which requires ATP.

D: *synaptic vesicles* contain neurotransmitters, and their exocytosis at a nerve terminus is an ATP-dependent process triggered by an action potential propagating along the neuron.

Vesicle fusion requires Ca^{2+} to enter the cell upon axon depolarization reaching the terminus.

38. C is correct.

p53 is a *tumor suppressor protein* that regulates the cell cycle and prevents cancer in multicellular organisms.

p53 is *the guardian of the genome* because of its role in conserving genetic stability by inhibiting genome mutations. The name p53 refers to its apparent molecular mass of 53 kDa.

Dalton is defined as 1/12 the mass of carbon and is a unit of a convention for expressing (i.e., often in kilodaltons or kDa) the molecular mass of proteins.

39. D is correct.

Newly synthesized secretory protein pathway:

rough ER → Golgi → secretory vesicles → exterior of the cell (via exocytosis)

Peroxisomes and *lysosomes* are destinations for proteins but do not involve secretory path exocytosis.

Ribosomes synthesize proteins, but the Golgi is the final organelle before exocytosis.

40. B is correct.

Plant cell membranes have higher amounts of unsaturated fatty acids.

The ratio of saturated and unsaturated fatty acids determines membrane fluidity.

Unsaturated fatty acids have kinks in their tails (due to double bonds) that push the phospholipids apart so the membrane retains its fluidity.

A: some textbooks state that plant membranes lack cholesterol, but a small amount is present, which is negligible compared to animal cells.

Cholesterol is usually dispersed in varying amounts throughout animal cell membranes in the irregular spaces between the hydrophobic lipid tails of the membrane.

Cholesterol functions as a *bidirectional buffer.*

At *elevated temperatures*, cholesterol *decreases membrane fluidity* because it confers stiffening and strengthening effects on the membrane.

At *low temperatures*, cholesterol *intercalates* between the phospholipids, preventing clustering and *stiffening the membrane.*

41. D is correct.

Nucleus, chloroplast, and mitochondria are organelles enclosed in a double membrane.

Endosymbiotic theory illustrates the evolution of mitochondria and chloroplasts from prokaryotes, thus explaining the resemblance.

42. C is correct.

Golgi apparatus processes *secretory proteins* via post-translational modifications.

Proteins targeted to the Golgi (from the rough ER) have three destinations:

1) secreted out of the cell,

2) transported into organelles, or

3) targeted to the plasma membrane (as a receptor, channel, or pore).

43. C is correct.

Carbon, hydrogen, and oxygen are in macromolecules (i.e., proteins, carbohydrates, nucleic acids, and lipids).

Nitrogen is in nucleic acids and proteins, which may contain sulfur or phosphorus.

Amino acids are *monomers* (building blocks) of proteins, and *nitrogen* is an *amino acid* and *urea* component.

Organic nitrogen is a nitrogen compound originating from a living organism.

Carbohydrates and lipids are made of only carbon, hydrogen, and oxygen

Carbohydrates have about 2 H and 1 O atom for every C atom.

Proteins are composed of C, H, O, N, and sometimes S.

Nucleic Acids are composed of: nucleotides composed of C, H, O, N, and P.

44. A is correct.

Albumin is the most abundant plasma protein; primarily responsible for *osmotic pressure* in circulatory systems.

Albumin is too large to pass from the circulatory system into the interstitial space.

Osmotic pressure is the force of H_2O flowing from an area of a lower solute to a higher solute concentration.

If a membrane is impermeable to a solute, H_2O flows across the membrane (i.e., osmosis) until the differences in solute concentrations have equilibrated.

Capillaries are impermeable to albumin (i.e., solute).

Increasing albumin concentration in the arteries and capillaries increases the movement of H_2O from the interstitial fluid to reduce the osmotic pressure in the arteries and capillaries.

45. B is correct.

Retinoblastoma protein (i.e., pRb or *RB1*) is a dysfunctional tumor suppressor protein.

pRb inhibits excessive cell growth by regulating the cell cycle through G1 (first gap phase) into S (DNA synthesis) phase.

pRb recruits chromatin-remodeling enzymes, such as methylases and acetylases.

46. E is correct.

Glycolysis occurs in the *cytoplasm*.

Krebs (TCA) cycle and *pyruvate oxidation* to acetyl-CoA occurs in the mitochondrial *matrix*.

Electron transport chain (ETC) uses cytochromes in the inner mitochondrial membrane and a proton (H^+) gradient in the intermembrane space).

A: *pyruvate* is oxidized to acetyl-CoA and transported from the cytoplasm into the matrix before joining oxaloacetate in the Krebs cycle.

B: Krebs cycle is the second stage of cellular respiration and occurs in the matrix of the mitochondrion.

C: *electron transport chain* is the final stage of cellular respiration occurring in the inner membrane (i.e., cytochromes) / intermembrane space (i.e., H^+ proton gradient) of the mitochondrion.

D: *reduction* (i.e., a gain of electrons) of FADH into $FADH_2$ occurs during the Krebs cycle.

47. D is correct.

Triglycerides are derived from glycerol and three fatty acids, commonly called "fats" and lipids, and are the most abundant lipid.

Teichoic acids are bacterial copolymers of carbohydrates and phosphate.

Peptidoglycan is formed by monosaccharide and amino acid polymers in bacterial cell walls.

Glycogen is a polysaccharide of glucose, a monosaccharide.

48. E is correct.

Secretory sequence in the flow of newly synthesized protein for export from the cell is:

$$\text{rough ER} \rightarrow \text{Golgi} \rightarrow \text{plasma membrane}$$

49. B is correct.

Minerals (e.g., potassium, sodium, calcium, and magnesium) are essential nutrients because they must be consumed in the diet and function as cofactors (i.e., nonorganic components) for enzymes.

Lysosomes do not digest minerals.

Organic molecules, such as nucleotides, proteins, and lipids, are hydrolyzed (i.e., degraded) into monomers.

Nucleotides have phosphate, sugar, and base; proteins have amino acids; lipids have glycerol and fatty acids.

50. D is correct.

Recycling of organelles within the cell is accomplished through autophagy by *lysosomes*.

51. B is correct.

Albumins are globular proteins in the circulatory system.

Carotenoids are organic pigments in chloroplasts of plants and photosynthetic organisms, such as some bacteria and fungi.

Carotenoids are fatty acid-like carbon chains containing *conjugated double bonds* and sometimes have six-membered carbon rings at ends.

Under visible light, they produce red, yellow, orange, and brown colors in plants and animals as pigments.

Waxes (esters of fatty acids and alcohols) are protective coatings on the skin, fur, leaves of higher plants, and on the exoskeleton cuticle of many insects.

Steroids (e.g., cholesterol, estrogen) have three fused cyclohexane rings and one cyclopentane ring.

Lecithin is an example of a phospholipid that contains glycerol, two fatty acids, a phosphate group, and nitrogen-containing alcohol.

52. A is correct.

Defective attachment of a chromosome to the spindle blocks activation of the *anaphase-promoting complex*.

Spindle checkpoint prevents anaphase onset in mitosis and meiosis until all chromosomes are attached to the spindle with the proper bipolar orientation.

53. B is correct.

Human gametes, formed during meiosis, are cells with a single copy (1N) of the genome.

After the second meiotic division, cells have a single unreplicated copy (i.e., devoid of a sister chromatid).

54. D is correct.

Smooth endoplasmic reticulum is the organelle for fatty acid, phospholipid, and steroid synthesis.

Smooth endoplasmic reticulum (SER) synthesizes lipids, phospholipids, and steroids; metabolism of carbohydrates and steroids; detoxification of alcohol and drugs; regulates Ca^{2+} concentration in muscle cells.

Endosome is a transport pathway starting at the Golgi.

Peroxisome undergoes beta-oxidation of long-chain fatty acids, which eventually yields CO_2 + H_2O.

Rough endoplasmic reticulum (RER) assembles proteins.

55. C is correct.

Centrioles are cylindrical structures mainly of *tubulin* in eukaryotic cells (except flowering plants and fungi).

Centrioles participate in the organization of the *mitotic spindle* and the completion of *cytokinesis*. They contribute to the structure of centrosomes and organize microtubules in the cytoplasm.

Centriole position determines the location of the nucleus and is crucial in the spatial arrangement of cells.

Plastid (e.g., chloroplast and chromoplast) are major organelles in plants and algae.

Plastids are the site of manufacturing and storing critical chemical compounds used by cells.

They often contain pigments used in photosynthesis. Pigments change to determine cell color.

Plastids, like prokaryotes, contain a circular double-stranded DNA molecule.

56. E is correct.

Centrioles are the organizational sites for microtubules (i.e., spindle fibers) that assemble during cell division (e.g., mitosis and meiosis).

Four phases of mitosis are *prophase, metaphase, anaphase,* and *telophase*.

Mitotic phases are followed by cytokinesis, which physically divides cells into two identical daughter cells.

Condensed chromosomes align along the *equatorial plane* in metaphase before the centromere (i.e., heterochromatin region of DNA) splits.

The two sister chromosomes begin their journey to the respective poles of the cell.

57. A is correct.

cAMP is a second messenger triggered when a ligand (e.g., peptide hormone or neurotransmitter) binds a membrane-bound receptor.

Adenylate cyclase (enzyme) is activated through a G-protein intermediate and converts ATP into cAMP.

Adenylate cyclase is attached to the inner layer of the phospholipid bilayer and is not in the cytoplasm.

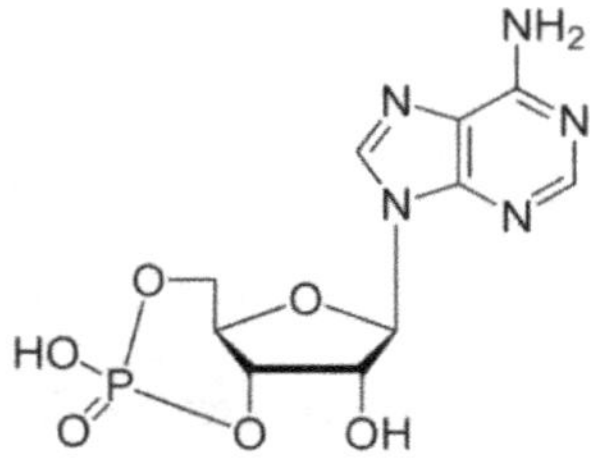

cAMP

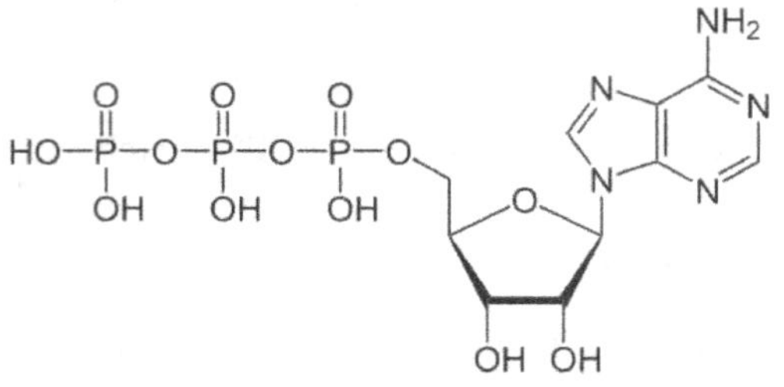

ATP

58. C is correct.

Cyclin and its associated *protein kinase* ensure a proper progression of cell division.

Cyclins are phosphorylated proteins responsible for specific events during cycle division (e.g., microtubule formation, chromatin remodeling).

Cyclins divide into *four classes* based on their behavior in the cell cycle: G1/S, S, M, and G1 cyclins.

59. A is correct.

Hypertonic solution is when there is a higher concentration of solutes outside the cell than inside.

When a cell is in a *hypertonic solution*, the tendency is for *water to flow out* of the cell to balance the concentration of the solutes.

Osmotic pressure draws water out of the cell, and the cell shrivels (i.e., crenation).

60. E is correct.

Peroxisomes are organelles abundant in the liver containing oxidases and detoxifying substances (i.e., alcohol and hydrogen peroxide).

61. B is correct.

Microtubules are hollow proteins composed of *tubulin* monomers necessary for:

1) formation of the *spindle apparatus* that separates chromosomes during cell division,

2) synthesis of *cilia* and *flagella*, and

3) formation of the *cytoskeleton* (within the cytoplasm).

II: actin and myosin are contractile fibers in muscle cells composed of microfilaments (not microtubules).

62. D is correct.

If a cell (e.g., pancreatic exocrine cell) is producing substantial amounts of proteins (i.e., enzymes) for export, this involves the *rough endoplasmic reticulum* (RER).

Protein synthesis begins in the nucleus with mRNA transcription, translated into polypeptides with the RER.

Exported proteins are packaged and modified in the Golgi; this cell would have a prominent nucleolus for rRNA (i.e., ribosomal components) synthesis.

63. A is correct.

Mitosis is cell division when new somatic (i.e., body) cells are added to multicellular organisms as they grow, and tissues are repaired or replaced.

Mitosis does not produce genetic variations. A daughter cell is identical in chromosome number and genetic makeup to the parental cell.

Eukaryotes divide by mitosis. Mitosis distributes identical genetic material to two daughter cells.

Fidelity of DNA transmission between generations without dilution is remarkable.

64. E is correct.

Amoeboids move using pseudopodia as bulges of cytoplasm powered by the elongation of flexible microfilaments (not microtubules).

Microtubules are long, hollow cylinders of polymerized α- and β-tubulin dimers critical in cellular processes such as maintaining cell structure and forming the cytoskeleton with microfilaments and intermediate filaments.

Microtubules comprise the internal structure of *cilia* and *flagella*.

Microtubules provide platforms for intracellular transport and participate in cellular processes, including the movement of secretory vesicles, organelles, and intracellular substances.

Microtubules engage in cell division (i.e., mitosis and meiosis), including the formation of mitotic spindles that pull apart eukaryotic chromosomes.

65. B is correct.

Apoptosis is programmed cell death that occurs during fetal development and aging.

Synaptic cleft development, the formation of separate digits in the hand of a fetus, and tadpole tail reabsorption are examples of apoptosis during development.

Synthesis of the uterine lining is an *anabolic* process that involves mitosis (i.e., cell division).

66. A is correct.

A cell involved in active transport (e.g., intestinal epithelial cells) requires much ATP.

Therefore, many mitochondria are needed to meet cellular respiration needs (i.e., glucose $\rightarrow$ ATP).

B: elevated levels of *DNA synthesis* occur in cells using mitosis for rapid reproduction (i.e., skin cells).

C: elevated levels of *adenylate cyclase* (i.e., cAMP second messenger) are in the target cells of peptide hormones.

D: *polyribosomes* are in cells with a high protein synthesis level.

E: *lysosomes* are organelles with low pH that function to digest intracellular molecules.

67. E is correct.

Coat proteins, like clathrin, form small vesicles to transport molecules within and between cells.

Endocytosis and *exocytosis* of vesicles allow cells to transfer nutrients, import signaling receptors, mediate immune responses, and degrade cell debris after tissue inflammation.

Endocytosis is when cells internalize receptor-ligand complexes from the cell surface (e.g., cholesterol bound to its receptor in the clathrin-coated pits).

Receptor-ligand complexes cluster in clathrin-coated pits at the cell surface and pinch off the vesicles that join acidic vesicles (i.e., endosomes).

68. D is correct.

Cells respond to peptide hormones through biochemical reactions involving membrane receptors and kinase activation.

First and second messengers communicate responses between cells because the G protein links the first and second messengers.

cAMP is a second messenger triggered when a ligand (e.g., peptide hormone or neurotransmitter) binds a membrane-bound receptor.

Adenylate cyclase (enzyme) is activated through a G-protein intermediate and converts ATP into cAMP.

Adenylate cyclase is attached to the inner layer of the phospholipid bilayer and is not in the cytoplasm.

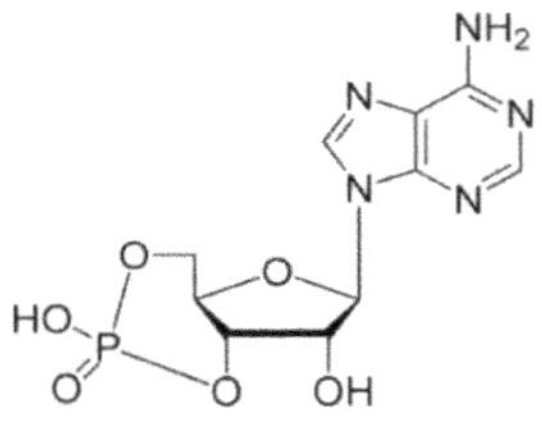

cAMP 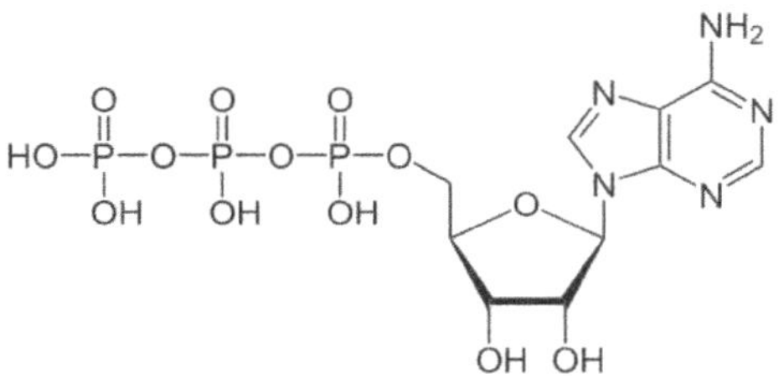   ATP

69. E is correct.

Transduction involves a virus and is one of three methods (*conjugation* and *transformation*) that bacterial cells use to introduce genetic variability into their genomes.

Transduction is when a virus introduces novel genetic material while infecting its host.

Mitosis occurs in somatic cells for growth and repair, producing two identical diploid (2N) daughter cells.

Notes for active learning

Notes for active learning

Notes for active learning

Practice Questions: Enzymes & Cellular Metabolism

1. The atom that generates a hydrogen bond to stabilize the α-helical configuration of a polypeptide is:

A. peptide bond atom

B. atom in the R-groups

C. hydrogen of carbonyl oxygen

D. hydrogen of the amino nitrogen

E. two of the above

2. The ATP molecule contains three phosphate groups, two of which are:

A. bound as phosphoanhydrides

B. bound to adenosine

C. never hydrolyzed from the molecule

D. cleaved off during most biochemical reactions

E. equivalent in energy for the hydrolysis of each of the phosphates

3. Which attractive force is used by side chains of nonpolar amino acids to interact with nonpolar amino acids?

A. ionic bonds

B. hydrogen bonds

C. hydrophobic interaction

D. disulfide bonds

E. dipole-dipole

4. Fermentation yields less energy than aerobic respiration because:

A. it requires a greater expenditure of cellular energy

B. glucose molecules are not completely oxidized

C. it requires more time for ATP production

D. oxaloacetic acid serves as the final H^+ acceptor

E. it occurs in H_2O

5. How would the reaction kinetics of an enzyme and its substrate change if an anti-substrate antibody is added?

A. The antibody binds to the substrate, which increases the V_{max}

B. The antibody binds to the substrate, which decreases K_m

C. No change because K_m and V_{max} are independent of antibody concentration

D. The antibody binds the substrate, which decreases V_{max}

E. The antibody binds to the substrate, which increases K_m

---

6. Metabolism is:

    A. consumption of energy

    B. release of energy

    C. all conversions of matter and energy taking place in an organism

    D. production of heat by chemical reactions

    E. exchange of nutrients and waste products with the environment

7. During alcoholic fermentation, all occur EXCEPT:

    A. release of CO_2

    B. oxidation of glyceraldehyde-3-phosphate

    C. oxygen is not consumed in the reaction

    D. ATP synthesis as a result of oxidative phosphorylation

    E. NAD^+ is produced

8. When determining a protein's amino acid sequence, acid hydrolysis causes partial destruction of tryptophan, conversion of asparagine into aspartic acid, and conversion of glutamine into glutamic acid. Which statement is NOT correct?

    A. Glutamine concentration is related to the level of aspartic acid

    B. Glutamic acid levels are an indirect indicator of glutamine concentration

    C. Tryptophan levels cannot be estimated accurately

    D. Asparagine levels cannot be estimated accurately

    E. Tryptophan and asparagine levels cannot be estimated accurately

9. What is the correct sequence of energy sources used by the body?

    A. fats → glucose → other carbohydrates → proteins

    B. glucose → other carbohydrates → fats → proteins

    C. glucose → other carbohydrates → proteins → fats

    D. glucose → fats → proteins → other carbohydrates

    E. fats → proteins → glucose → other carbohydrates

10. Enzymes act by:

    A. lowering the overall free energy change of the reaction

    B. decreasing the distance reactants must diffuse to find each other

    C. increasing the activation energy

    D. shifting equilibrium towards product formation

    E. decreasing the activation energy

11. For the following reaction, which statement is TRUE?

$$ATP + Glucose \rightarrow Glucose\text{-}6\text{-}phosphate + ADP$$

A. reaction results in the formation of a phosphodiester bond
B. reaction is endergonic
C. reaction is part of the Krebs cycle
D. free energy change for the reaction is approx. –4 kcal
E. reaction does not require an enzyme

12. Which is the correct cAMP classification because cAMP-dependent protein phosphorylation activates hormone-sensitive lipase?

A. DNA polymerase
B. lipoproteins

C. glycosphingolipids
D. second messenger
E. phospholipids

13. Which statement is NOT true about the Krebs cycle?

A. Krebs cycle occurs in the matrix of the mitochondria
B. Citrate is an intermediate in the Krebs cycle
C. Krebs cycle produces nucleotides such as NADH and $FADH_2$
D. Krebs cycle is linked to glycolysis by pyruvate
E. Krebs cycle is the single greatest direct source of ATP in the cell

14. The rate of V_{max} is directly related to:

I. Enzyme concentration
II. Substrate concentration
III. Concentration of a competitive inhibitor

A. I, II, and III
B. I and III only

C. I and II only
D. I only
E. II and III only

15. A reaction in which the substrate glucose binds to the enzyme hexokinase, and the conformation of both molecules changes, is an example of:

A. lock-and-key mechanism
B. induced-fit mechanism
C. competitive inhibition
D. allosteric inhibition
E. covalent bond formation at the active site

16. You are studying an enzyme that catalyzes a reaction with a free energy change of +5 kcal. If you double the amount of enzyme in a reaction mixture, what would be the free energy change for the reaction?

A. −10 kcal

B. −5 kcal

C. 0 kcal

D. +5 kcal

E. +10 kcal

17. Glucokinase and hexokinase catalyze the first glycolysis reaction; glucokinase has a higher K_m. Which is a correct statement if K_m is equal to [Substrate] = $1/2 V_{max}$?

A. hexokinase is always functional and is not regulated by negative feedback

B. hexokinase and glucokinase are not isozymes

C. glucokinase is not a zymogen

D. glucokinase becomes active from elevated levels of fructose

E. none of the above

18. α-helices and β-pleated sheets are characteristic of which level of protein folding?

A. primary

B. secondary

C. tertiary

D. quaternary

E. secondary & tertiary

19. Coenzymes are:

A. minerals such as Ca^{2+} and Mg^{2+}

B. small inorganic molecules that work with an enzyme to enhance the reaction rate

C. linking together of two or more enzymes

D. small molecules that do not regulate enzymes

E. small organic molecules that work with an enzyme to enhance the reaction rate

20. Hemoglobin is an example of a protein that:

A. is initially inactive in cell

B. has a quaternary structure

C. conducts a catalytic reaction

D. has only a tertiary structure

E. has a signal sequence

21. The site of the TCA cycle in eukaryotic cells, as opposed to prokaryotes, is:

   A. mitochondria

   B. endoplasmic reticulum

   C. cytosol

   D. nucleolus

   E. intermembrane space of the mitochondria

22. All are metabolic waste products, EXCEPT:

   A. lactate   **C.** CO_2

   B. pyruvate   **D.** H_2O

                                         E. ammonia

23. When measuring the reaction velocity as a function of substrate concentration, what is likely to occur if the enzyme concentration changes?

   A. V_{max} changes, while K_m remains constant

   B. V_{max} remains constant, while V changes

   C. V_{max} remains constant, while K_m changes

   D. V_{max} remains constant, while V and K_m change

   E. Not possible to predict without experimental data

24. A holoenzyme is:

   A. inactive enzyme without its cofactor

   B. inactive enzyme without its coenzyme

   C. active enzyme with its organic moiety

   D. active enzyme with its coenzyme

   E. active enzyme with its cofactor

25. *Clostridium butyricum* is a heterotrophic anaerobe that grows on glucose, converting it to butyric acid as a product. If the free energy for this reaction is –50 kcal, the maximum number of ATP that this organism can synthesize from one molecule of glucose is approximately:

   A. 5 ATP   **C.** 7 ATP

   B. 36 ATP   **D.** 38 ATP

                                       E. 0 ATP

26. Which amino acid is directly affected by dithiothreitol (DTT), known to reduce and break disulfide bonds?

A. methionine

B. leucine

C. glutamine

D. cysteine

E. proline

27. Which answer represents a correct pairing of aspects for cellular respiration?

A. Krebs cycle – cytoplasm

B. fatty acid degradation – lysosomes

C. electron transport chain – inner mitochondrial membrane

D. glycolysis – inner mitochondrial membrane

E. ATP synthesis – outer mitochondrial membrane

28. An apoenzyme is:

A. active enzyme with its organic moiety

B. inactive enzyme without its inorganic cofactor

C. active enzyme with its cofactor

D. active enzyme with its coenzyme

E. inactive enzyme without its cofactor

29. Several forces stabilize tertiary structure of a protein. Which is likely involved in this stabilization?

A. glycosidic bonds

B. disulfide bonds

C. peptide bonds

D. anhydride bonds

E. phosphodiester bonds

30. In the non-oxidative branch of the pentose phosphate pathway, transketolase is a reaction catalyst enzyme, and its activity depends on a prosthetic group. Which bond is used by a prosthetic group to attach to its target?

A. van der Waals interactions

B. covalent bond

C. ionic bond

D. hydrogen bond

E. dipole-dipole interactions

31. The process of $C_6H_{12}O_6 + O_2 \rightarrow CO_2 + H_2O$ is completed in the:

A. plasma membrane

B. cytoplasm

C. ribosome

D. nucleus

E. mitochondria

32. In hyperthyroidism, oxidative metabolism rates measured through basal metabolic rate (BMR) will be:

A. indeterminable

B. below normal

C. above normal

D. normal

E. between below normal to normal

33. Cofactors are:

    I. small inorganic molecules that work with an enzyme to enhance the reaction rate

    II. small organic molecules that work with an enzyme to enhance the reaction rate

    III. small molecules that regulate enzyme activity

A. I only

B. I and II only

C. II and III only

D. I, II and III

E. I and III only

34. All proteins:

A. are post-translationally modified

B. have a primary structure

C. have catalytic activity

D. contain prosthetic groups

E. contain disulfide bonds

35. After being gently denatured with the denaturant removed, proteins can recover significant activity because recovery of the structure depends on?

A. 4° structure of the polypeptide

B. 3° structure of the polypeptide

C. 2° structure of the polypeptide

D. 1° structure of the polypeptide

E. interactions between the polypeptide and its prosthetic groups

36. All statements about glycolysis are true, EXCEPT:

    A. end-product can be lactate, ethanol, CO_2, and pyruvate

    B. $FADH_2$ is produced during glycolysis

    C. a molecule of glucose is converted into two molecules of pyruvate

    D. net total of two ATPs are produced

    E. NADH is produced

37. Vitamins are:

    A. necessary components in the human diet

    B. present in plants but not in animals

    C. absent in bacteria within the gastrointestinal tract

    D. all water-soluble

    E. inorganic components of the diet

38. Hemoglobin is a protein that contains a:

    A. site where proteolysis occurs   **C.** bound zinc atom

    B. phosphate group at its active site   **D.** prosthetic group

                                              E. serine phosphate at its active site

39. Which amino acid is nonoptically active, lacking four different groups bonded to the α carbon?

    A. valine   **C.** glutamate

    B. aspartic acid   **D.** cysteine

                                      E. glycine

40. Which statement is TRUE for the glycolytic pathway?

    A. glucose produces a net of two molecules of ATP and two molecules of NADH

    B. glucose produces one molecule of pyruvate

    C. O_2 is a reactant for glycolysis

    D. glucose is partially reduced

    E. pyruvate is the final product of the Krebs cycle and is intermediate for the next series of reactions in cellular respiration

41. Which metabolic process occurs in the mitochondria?

I. Krebs cycle   II. glycolysis   III. electron transport chain

A. II only
B. I and III only

C. II and III only
D. I, II and III
E. I and II only

42. Which statement below best describes the usual relationship of the inhibitor molecule to the allosteric enzyme in feedback inhibition of enzyme activity?

A. The inhibitor is the substrate of the enzyme
B. The inhibitor is the product of the enzyme-catalyzed reaction
C. The inhibitor is the final product of the metabolic pathway
D. The inhibitor is a metabolically unrelated signal molecule
E. The inhibitor binds to a tertiary protein

43. *Clostridium butyricum* is an obligate anaerobe that grows on glucose and converts it to butyric acid. If the ΔG for this reaction is -50 kcal, the synthesis of ATP occurs through:

A. substrate-level phosphorylation
B. oxidative phosphorylation
C. neither substrate-level nor oxidative phosphorylation
D. both substrate-level and oxidative phosphorylation
E. electron transport cascade

44. While covalent bonds are the strongest bonds of protein structure, which are connected by a peptide bond?

A. ammonium group and ester group
B. two amino groups

C. the α carbons
D. two carboxylate groups
E. amino group and carboxylate group

45. All statements apply to oxidative phosphorylation, EXCEPT:

A. it can occur under anaerobic conditions
B. it produces two ATPs for each $FADH_2$
C. it involves O_2 as the final electron acceptor
D. it occurs on the inner membrane of the mitochondrion
E. it involves a cytochrome electron transport chain

46. Like other catalysts, enzymes:

    I. increase the rate of reactions without affecting ΔG

    II. shift the chemical equilibrium from more reactants to more products

    III. do not alter the chemical equilibrium between reactants and products

A. I only

B. I and II only

C. I and III only

D. III only

E. II and III only

47. Enzyme activity can be regulated by:

    I. zymogen proteolysis

    II. changes in substrate concentration

    III. post-translational modifications

A. I only

B. II and III only

C. I, II and III

D. I and II only

E. I and III only

48. Which interactions stabilize parallel and non-parallel beta-pleated sheets?

A. hydrophobic interactions

B. hydrogen bonds

C. van der Waals interactions

D. covalent bonds

E. dipole-dipole interactions

49. Glycogen is:

A. degraded by glycogenesis

B. synthesized by glycogenolysis

C. unbranched molecule

D. in both plants and animals

E. the storage polymer of glucose

50. If $[S] = 2\ K_m$, what portion of active sites of the enzyme is filled by substrate?

A. 3/4

B. 2/3

C. 1/2

D. 1/3

E. 1/4

51. In allosteric regulation, how is enzyme activity affected by binding a small regulatory molecule to the enzyme?

A. It is inhibited
B. It is stimulated
C. Can be either stimulated or inhibited
D. Is neither stimulated nor inhibited
E. The rate increases to twice K_m and then plateaus

52. All biological reactions:

A. are exergonic
B. have an activation energy

C. are endergonic
D. occur without a catalyst
E. are irreversible

53. In eukaryotes, energy is trapped in a high-energy phosphate group during oxidative phosphorylation in the:

A. nucleus
B. mitochondrial matrix
C. inner mitochondrial membrane
D. outer mitochondrial membrane
E. cytoplasmic face of the plasma membrane

54. All statements about enzymes are true, EXCEPT:

A. They function optimally at a particular temperature
B. They function optimally at a particular pH
C. They may interact with non-protein molecules to achieve biological activity
D. Their activity is not affected by a genetic mutation
E. They are almost always proteins

55. The Gibbs free-energy change (ΔG) of a reaction is determined by:

I. intrinsic properties of the reactants and products
II. concentrations of the reactants and products
III. the temperature of the reactants and products

A. I only
B. I and III only

C. II and III only
D. I, II and III
E. I and II only

56. Allosteric enzymes:

    I. are regulated by metabolites that bind at sites other than the active site

    II. have quaternary structure

    III. show cooperative binding of substrate

A. I only

B. I and II only

C. II and III only

D. I, II and III

E. III only

57. Which symptom is characteristic of a patient exposed to monoamine oxidase inhibitors, given that they prevent the breakdown of catecholamines (i.e., epinephrine)?

A. decreased blood flow to skeletal muscles

B. excessive digestive activity

C. dilated pupils

D. decreased heart rate

E. increased peristalsis along the GI tract

58. The ΔG for hydrolysis of ATP to ADP and Pi is:

A. Greater than +7.3 kcal/mole

B. +7.3 kcal/mole

C. −7.3 kcal/mole

D. −0.5 kcal/mole

E. none of the above

59. The active site of an enzyme is where:

    I. prosthetic group is bound

    II. proteolysis occurs for zymogens

    III. non-competitive inhibitors bind

A. I only

B. II only

C. I and II only

D. II and III only

E. I, II and III

60. The hydrolysis of ATP $\rightarrow$ ADP + phosphate allows glucose-6-phosphate to be synthesized from glucose and phosphate because:

    A. heat produced from ATP hydrolysis drives the glucose-6-phosphate synthesis
    B. enzymatic coupling of these two reactions allows the energy of ATP hydrolysis to drive the synthesis of glucose-6-phosphate
    C. energy of glucose phosphorylation drives ATP splitting
    D. all the above
    E. None of the above

61. Which statement is TRUE?

    A. Protein function can be altered by modification after synthesis
    B. Covalent bonds stabilize the secondary structure of proteins
    C. A protein with a single polypeptide subunit has a quaternary structure
    D. Integral membrane proteins contain high amounts of acidic amino acids
    E. Protein denaturation is always reversible

62. What does alcoholic fermentation have in common with pyruvate oxidation under aerobic conditions?

    A. no commonality
    B. triose sugar is a product of each reaction
    C. ethyl alcohol is a product of each reaction
    D. CO_2 is a product of each reaction
    E. NADH is a product of each reaction

63. Which statement is TRUE?

    A. All disaccharides must contain fructose
    B. Most polysaccharides contain ribose
    C. All polysaccharides are energy-generating molecules
    D. Polysaccharides are only in animal cells
    E. Glucose and fructose have different chemical properties even with the same molecular formula

Notes or active learning

Detailed Explanations: Enzymes & Cellular Metabolism

Answer Key

1: D	11: A	21: A	31: E	41: B	51: C	61: A
2: A	12: D	22: B	32: C	42: C	52: B	62: D
3: C	13: E	23: A	33: D	43: A	53: C	63: E
4: B	14: D	24: E	34: B	44: E	54: D	
5: E	15: B	25: C	35: D	45: A	55: D	
6: C	16: D	26: D	36: B	46: C	56: D	
7: D	17: C	27: C	37: A	47: C	57: C	
8: A	18: B	28: E	38: D	48: B	58: C	
9: B	19: E	29: B	39: E	49: E	59: C	
10: E	20: B	30: B	40: A	50: B	60: B	

1. D is correct.

α-helix (alpha helix) is a typical secondary structure of proteins.

α-helix is a *right-handed coiled* or spiral conformation (helix) when each backbone amino (N-H) group donates partial positive hydrogen to form a hydrogen bond with a lone pair of electrons from the backbone carbonyl (C=O) group of another amino acid four residues earlier.

2. A is correct.

ATP is a nucleotide composed of adenosine (A), ribose, and three phosphate groups.

3. C is correct.

Hydrophobic side chain groups (e.g., phenylalanine, methionine, leucine, and valine) interact through hydrophobic interactions, excluding water from the attraction region.

Nine amino acids with hydrophobic side chains are alanine (Ala), glycine (Gly), isoleucine (Ile), leucine (Leu), methionine (Met), phenylalanine (Phe), proline (Pro), tryptophan (Trp), and valine (Val).

B: *hydrogen bonds* require hydrogen to be connected to an electronegative atom of fluorine, nitrogen, oxygen, or chlorine (weaker hydrogen bond due to the larger valence shell size) to establish a δ+ and δ- (partial positive and negative regions that attract).

D: *cysteine* is the only amino acid capable of forming disulfide bonds because the side chain contains sulfur.

E: *dipole-dipole interactions* (collectively Van der Waals) are weak interactions from the close association of permanent or induced dipoles.

Proteins contain many dipole interactions, which vary considerably in strength.

4. B is correct.

Cellular respiration uses enzyme-catalyzed reactions as a catabolic pathway using potential energy of glucose.

While glycolysis yields two ATP per glucose, cellular respiration (i.e., glycolysis, Krebs/TCA cycle, and electron transport chain) produces about 30–32 ATP.

Around 2004, research reduced the estimate of 36-38 ATP due to inefficiencies of ATP synthase.

Cellular respiration is an aerobic process whereby O_2 is the final acceptor of electrons passed from cytochrome electron carriers during the final stage of the electron transport chain.

Glycolysis (glucose to pyruvate) occurs in the cytoplasm.

Krebs cycle (TCA) occurs in the matrix (i.e., the cytoplasm of mitochondria), while the electron transport chain occurs in the intermembrane space of the mitochondria for eukaryotes.

For prokaryotes, both processes occur in the cytoplasm because prokaryotes lack mitochondria.

Some forms of anaerobic respiration use an electron acceptor other than O_2 (e.g., iron, cobalt, or manganese reduction) at the end of the electron transport chain.

Glucose molecules are entirely oxidized in these cases because glucose products enter the Krebs cycle.

Highly reduced chemical compounds (e.g., NADH or $FADH_2$) establish an electrochemical gradient across a membrane in the Krebs cycle.

The reduced chemical compounds (NADH or $FADH_2$) are oxidized by integral membrane proteins that transfer the electron to the final electron acceptor.

5. E is correct.

K_m is the substrate concentration [S] at half the maximum reaction velocity (V_{max}) and increases when the antibody binds the substrate.

This reaction kinetics follow the profile of a *competitive inhibitor* (visualized by the y-axis intercept) in the Lineweaver-Burke (double reciprocal).

The substrate binds the antibody and is not available to react with the enzyme; therefore, an additional substrate is needed to bind the same amount of enzyme (as a lower substrate concentration in the absence of an antibody).

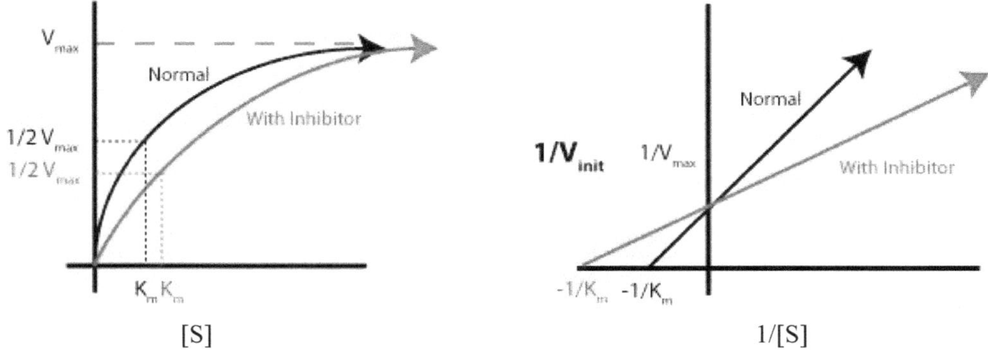

Michaelis-Menten equation (left) and Lineweaver-Burke (double reciprocal) plot

Graph on *left* illustrates the Michaelis-Menten equation for enzyme kinetics.

Graph on *right* is the Lineweaver-Burke (double reciprocal) plot showing the change in K_m when the substrate is subjected to competitive inhibition.

V_{max} does not change when the number of available substrates changes.

Binding to a substrate does not affect V_{max} but affects K_m because the amount of available substrate is reduced, altering the reaction kinetics because it binds the substrate.

6. C is correct.

Metabolism means change and refers to life-sustaining chemical transformations within cells.

Enzyme-catalyzed reactions allow organisms to maintain structures, grow, reproduce, and respond.

Metabolism includes *anabolism* (or *buildup*) and *catabolism* (or *breakdown*).

Catabolism is the breakdown of organic matter and harvests energy from cellular respiration (e.g., glycolysis, Krebs cycle, and the electron transport chain).

Anabolism utilizes energy (e.g., ATP) to *synthesize biomolecules* (e.g., lipids, nucleic acids, proteins).

7. D is correct.

Fermentation is an anaerobic process and occurs in the absence of oxygen.

Purpose of fermentation is to *regenerate* the high-energy nucleotide intermediate of NAD^+.

Oxidative phosphorylation occurs during the electron transport chain as the final stage of cellular respiration.

8. A is correct.

Acid hydrolysis has several effects:

1) partial destruction of tryptophan prevents the proper estimate of the tryptophan concentration;

2) conversion of asparagine into aspartic acid prevents the direct measure of asparagine;

3) conversion of glutamine to glutamic acid.

Therefore, the concentration of glutamic acid is an indirect measure of glutamine.

9. B is correct.

When plasma glucose levels are low, the body utilizes other energy sources for cellular metabolism.

Energy sources are used in preferential order:

glucose → other carbohydrates → fats → proteins

These molecules are converted to glucose or glucose intermediates degraded in the glycolytic pathway and the Krebs cycle (i.e., citric acid cycle).

Proteins are used last for energy because there is no protein storage in the body.

Catabolism of protein results in muscle wasting and connective tissue breakdown, which is harmful over time.

10. E is correct.

Activation energy is the minimum amount of energy needed for a reaction. It is considered an "energy barrier" because energy must be added to the system.

Enzymes act by *decreasing the activation energy* (or *lowering the energy barrier*), allowing the reaction, and increasing the reaction rate.

11. A is correct.

Phosphate group from ATP is transferred to glucose, creating a *phosphodiester bond.*

$$\text{ATP} + \text{Glucose} \rightarrow \text{Glucose-6-phosphate} + \text{ADP}$$

12. D is correct.

Second messengers are common in eukaryotes.

Cyclic nucleotides (e.g., cAMP and cGMP) are second messengers transmitting extracellular signals from the cell membrane to intracellular proteins.

Second messengers are used by hydrophilic protein hormones (e.g., insulin, adrenaline) that cannot cross the plasma membrane.

Steroid hormones (e.g., testosterone, estrogen, progesterone) are lipid-soluble and can pass through the plasma membrane and enter the cytoplasm without second messengers.

13. E is correct.

Greatest direct source of ATP synthesis involves the *electron transport chain.*

Glycolysis occurs in the cytoplasm, the *Krebs cycle* (TCA) in the matrix of mitochondria, and *oxidative phosphorylation* (i.e., the electron transport chain) in the intermembrane space and inner membrane of mitochondria.

Glycolysis produces pyruvate, converted into acetyl CoA, and joined to oxaloacetate in the Krebs cycle, producing citrate as the first intermediate.

Krebs cycle produces two GTP (i.e., ATP) per glucose molecule (or 1 GTP per pyruvate).

Most ATP is formed by *oxidative phosphorylation* when NADH and $FADH_2$ nucleotides are oxidized and donate their electrons to the cytochromes in the electron transport chain.

14. D is correct.

V_{max} is the reaction velocity at a fixed enzyme concentration and depends on the total enzyme concentration. Adding enzymes allows more enzyme reactions per minute.

II: V, not V_{max}, depends on [S]. V_{max} is a constant for a specified amount of enzyme.

III: *competitive inhibition* is when the inhibitor binds reversibly to the active site. Adding enough substrate overcomes competitive inhibition, and the original V_{max} is obtained.

15. B is correct.

Two major models of *enzyme-substrate binding*:

> *Induced-fit model* is when the initial interactions between enzyme and substrate (e.g., hexokinase and glucose) are weak, but the weak interactions induce *conformational changes* that strengthen binding.

> *Lock-and-key* model is when the molecules conform for binding and do not change.

16. D is correct.

Gibbs free energy (ΔG) of a reaction is not dependent on the amount of enzyme.

Gibbs free energy measures the energy difference (Δ) between the reactants and products.

> ΔG is *positive* for endergonic reactions with products higher in energy than the reactants (absorbs energy).

> ΔG is *negative* for exergonic reactions when products are lower in energy than the reactants (releases energy).

17. C is correct.

Zymogens are inactive forms of enzymes and have an ~ogen suffix (e.g., pepsinogen).

From the question, glucokinase has a higher Michaelis constant (K_m).

A higher value for K_m is due to enzymes having a lower affinity for the reactant (e.g., glucose).

Information about K_m does not answer the question and is merely a distraction.

A: *hexokinase* is a control point enzyme (i.e., irreversible steps) regulated by negative feedback inhibition.

B: *hexokinase* and glucokinase catalyze the same reaction (from the question); by definition, *isozymes*.

D: *fructose* is not a reactant in glycolysis.

18. B is correct.

Secondary structure (2°) is the repetition of α-helices or β-pleated sheets in the polypeptide backbone.

Primary structure (1°) is the linear sequence of amino acids.

Tertiary structure (3°) involves interactions between the side chains of the amino acids.

Quaternary structure (4°) requires the interaction of two or more polypeptides. 4° requires more than one polypeptide chain in the mature protein (e.g., hemoglobin).

19. E is correct.

Cofactors are organic or inorganic molecules classified depending on how tightly they bind to an enzyme.

Coenzymes are loosely bound *organic cofactors* released from the enzyme's active site during the reaction (e.g., ATP, NADH).

Prosthetic groups are tightly bound organic cofactors.

20. B is correct.

Quaternary (4°) *structure* of proteins involves two or more polypeptide chains.

Hydrophobic interactions and disulfide bridges between cysteines maintain the quaternary structure between the different polypeptide chains (e.g., 4 chains of 2 α and 2 β in hemoglobin).

21. A is correct.

Mitochondrion is the organelle bound by a double membrane, where the Krebs (TCA) cycle, electron transport, and oxidative phosphorylation occur.

In eukaryotes, the Krebs cycle (i.e., TCA) occurs in the matrix of the mitochondria.

Matrix is an interior space in the mitochondria analogous to the cell cytoplasm.

B: *smooth ER* (endoplasmic reticulum) is connected to the nuclear envelope and synthesizes lipids, phospholipids, and steroids.

Cells secreting lipids, phospholipids, and steroids (e.g., testes, ovaries, skin oil glands) have prominent smooth endoplasmic reticulum.

Smooth ER carries out the metabolism of carbohydrates, drug detoxification, attachment of receptors on cell membrane proteins, and steroid metabolism.

C: *cytosol* is common to eukaryotes and prokaryotes and equivalent to the cytoplasm.

D: *nucleolus* is the organelle within the nucleus responsible for synthesizing ribosomal RNA (rRNA).

E: *intermembrane space of the mitochondria* is where the electron transport chain (ETC) establishes a proton gradient (or chemiosmotic gradient) with concentrated H^+ ions forced into the space in eukaryotes.

22. B is correct.

Pyruvate is the product of glycolysis, converted into acetyl Co-A as the starting reactant for the Krebs cycle.

Pyruvate is *not* a waste product of cellular respiration. Metabolic waste products are from enzymatic processes.

Pyruvate, made during glycolysis (as an intermediate of cellular respiration), is converted to acetyl CoA and enters the Krebs cycle to be oxidized into CO_2 and H_2O or converted into the *waste product of lactate* under anaerobic conditions. Waste products can damage cells and must be removed.

A: *lactate* (2 carbons) is produced from *pyruvate* (3-carbon chain) as the waste product of anaerobic respiration.

Lactate is converted to pyruvate in the liver when O_2 becomes available.

If lactate is not metabolized, it can lead to lactic acidosis (acidification of the blood) and death.

C and D: CO_2 and H_2O are waste products of aerobic respiration.

CO_2 and H_2O are removed from the lungs via expiration during regular breathing.

CO_2 increases lead to acidosis (lowering blood pH) because CO_2 reacts with H_2O to form carbonic acid (H_2CO_3).

E: *ammonia* is the waste product of protein metabolism, converted to the less toxic urea in the liver and removed by the kidneys.

If ammonia is not converted and cleared, the blood becomes alkaline (higher in pH), potentially fatal.

23. A is correct.

V_{max} is proportional to the number of active sites on the enzyme (i.e., [enzyme] or enzyme concentration).

K_m remains constant because it is a measure of the active site affinity for the substrate.

Regardless of the number of enzyme molecules, each enzyme interacts with the substrate similarly.

Models propose mechanisms of enzymatic catalysis.

Michaelis–Menten model describes enzyme-substrate interactions, whereby enzymatic catalysis occurs at a specific site on enzymes – the active site.

Substrate (S) binds the enzyme (E) at the active site for an enzyme-substrate (ES) complex.

ES complex dissociates into enzyme and substrate or moves forward in the reaction to form product (P) and original enzyme.

$$E + S \underset{k_{-1}}{\overset{k_1}{\rightleftharpoons}} ES \overset{k_2}{\rightarrow} P + E$$

Rate constants for different steps in the reaction are k_1, k_{-1}, and k_2.

The overall rate of product formation is a combination of each rate constant.

At constant enzyme concentrations, varying [S] changes the rate of product formation.

At low [S], a small fraction of enzyme molecules is occupied with the substrate.

ES and [P] formation rate increases linearly with substrate concentration.

At high [S], all active sites are occupied with the substrate, and increasing the substrate concentration further does not increase the reaction rate, and the reaction rate is V_{max}.

Relationship between substrate concentration and reaction rate:

$$V = \frac{[S]}{[S] + K_m} V_{max}$$

24. E is correct.

Apoenzyme, with its cofactors, is a *holoenzyme* (an active form).

Enzymes that require a cofactor, but do not have one bound, are *apoenzymes* (or apoproteins).

Cofactors can be inorganic (e.g., metal ions) or organic (e.g., vitamins).

Organic cofactors are coenzymes or prosthetic groups.

Most cofactors are not covalently attached to an enzyme but are tightly bound.

Organic prosthetic groups can be covalently bound.

Holoenzyme refers to enzymes containing multiple protein subunits (e.g., DNA polymerases), whereby the holoenzyme is the complete complex with all subunits needed for the activity.

---

25. C is correct.

ATP releases about -7.4 kcal/moles.

As an approximation, 1 kcal/mol $= \sim 4.2$ kJ/mol

Structure of the nucleotide of ATP (3 phosphates + ribose sugar + adenosine)

ATP contains three phosphate groups and is produced by enzymes (e.g., ATP synthase) from adenosine diphosphate (ADP) or adenosine monophosphate (AMP) and phosphate group donors.

Three central mechanisms of ATP biosynthesis are *oxidative phosphorylation* (in cellular respiration), *substrate-level phosphorylation*, and *photophosphorylation* (in photosynthesis).

Metabolic processes using ATP convert it back into its precursors; ATP is continuously recycled.

26. D is correct.

Cysteine is the only amino acid capable of forming a *disulfide* (S–S covalent) bond and serves an essential structural role in many proteins.

Thiol side chain in cysteine often participates in enzymatic reactions, serving as a nucleophile.

Thiol is susceptible to oxidization to give the *disulfide* derivative *cystine.*

27. C is correct.

Electron transport chain (ETC) uses cytochromes in the inner mitochondrial membrane and is a complex carrier mechanism of electrons that produces ATP through *oxidative phosphorylation.*

A: Krebs (i.e., TCA) cycle occurs in the *matrix* of the mitochondria.

Krebs cycle begins when acetyl CoA (i.e., 2-carbon chain) combines with oxaloacetic acid (i.e., 4-carbon chain) to form citrate (i.e., 6-carbon chain).

A series of additional enzyme-catalyzed reactions result in the oxidation of citrate and the release of two CO_2, and oxaloacetic acid is regenerated to begin the cycle again once it is joined to an incoming acetyl CoA.

B: *long-chain fatty acid degradation* occurs in peroxisomes that hydrolyze fat into smaller molecules fed into the Krebs cycle and used for cellular energy.

D: *glycolysis* occurs in the cytoplasm as the oxidative breakdown of glucose (i.e., 6-carbon chain) into two pyruvates (i.e., 3-carbon chain).

E: *ATP synthesis* occurs in the cytoplasm (glycolysis), the mitochondrial matrix (Krebs cycle), and the inner mitochondrial membrane (ETC).

28. E is correct.

Apoenzyme is an inactive enzyme without its cofactor.

Apoenzyme, with its cofactor(s), is a *holoenzyme* (an active form).

Enzymes that require a cofactor, but do not have one bound, are *apoenzymes* (or apoproteins).

Cofactors are inorganic (e.g., metal ions) or organic (e.g., vitamins).

Organic cofactors are *coenzymes* or *prosthetic groups*.

29. B is correct.

Disulfide bond is a covalent bond derived by coupling two thiols (~S-H) groups as R–S–S–R.

Peptide bonds join adjacent amino acids and make the protein's linear sequence (i.e., primary structure).

30. B is correct.

Enzyme prosthetic groups attach via strong molecular forces (e.g., covalent bonds).

A: *Van der Waals* interactions, like dipole/dipole, are weak molecular forces.

C: *ionic bonds* are powerful in non-aqueous solutions but are not used to attach prosthetic groups because they are weak in aqueous environments such as plasma. For example, NaCl dissociates in water.

D: *hydrogen bonds* (like hydrophobic interactions) are an example of a weak molecular force.

E: *dipole-dipole interactions* are an intermolecular force intermediate in strength between hydrogen bonds and van der Waals interactions that require a dipole (separation in charge) within the bonding molecules.

31. E is correct.

Cellular respiration begins via glycolysis in the cytoplasm but is completed in the mitochondria.

Glucose (six-carbon chain) is cleaved into two pyruvate molecules (three-carbon chain).

Pyruvate is converted to acetyl CoA that enters the Krebs cycle in the mitochondria.

NADH (from glycolysis and the Krebs cycle) and $FADH_2$ (from the Krebs cycle) enter the electron transport chain on the inner membrane of the mitochondria.

32. C is correct.

Primary activity of the thyroid hormone is to *increase* the *basal metabolic rate*.

33. D is correct.

Cofactors are organic or inorganic molecules classified depending on how tightly they bind to an enzyme.

Coenzymes are loosely bound organic cofactors released from the enzyme's active site during the reaction (e.g., ATP, NADH).

Prosthetic groups are tightly bound organic cofactors.

34. B is correct.

Primary structure is the *linear sequence* of amino acids in the polypeptide.

35. D is correct.

Denaturing the polypeptide disrupts the 4°, 3° and 2° structures, while the 1° structure (i.e., linear sequence of amino acids) is unchanged.

In folded proteins, 1° structure determines subsequent 2°, 3° and 4° structures.

36. B is correct.

$FADH_2$ is produced in the Krebs cycle (TCA) in cellular respiration.

A: product of glycolysis is pyruvate (i.e., pyruvic acid), which is converted to *lactic acid* (in humans) or *ethanol & CO_2* (in yeast) *via* anaerobic conditions.

Yeast is used for alcoholic production and baking for the dough to rise.

C: in glycolysis, the first series of aerobic (or anaerobic) reactions occur in the cytoplasm by breaking the 6-carbon glucose into two 3-carbon pyruvate molecules.

D: two ATP are required, and four ATP are produced = net of two ATP at the end of glycolysis.

E: two NADH are produced by glycolysis and are oxidized (i.e., lose electrons) to compound Q in the electron transport chain within the mitochondria.

37. A is correct.

There are 13 essential (i.e., must be consumed) vitamins: A, B_1 (thiamine), B_2 (riboflavin), B_3 (niacin), B_5 (pantothenic acid), B_6 (pyridoxine), B_7 (biotin), B_9 (folate) and B_{12} (cobalamin), C, D, E and K.

Four fat-soluble vitamins, A, D, E, and K are stored in adipose tissue.

38. D is correct.

Hemoglobin is a *quaternary protein* with two α and two β chains with Fe^{2+} (or Fe^{3+}) as the prosthetic group.

Prosthetic groups are *tightly bound* cofactors (i.e., nonorganic molecules such as metals).

39. E is correct.

Glycine is the only achiral amino acid because hydrogen is the R group (i.e., side chain).

Optical activity is the ability of a molecule to rotate plane-polarized light and requires a chiral center (i.e., carbon atom bound to 4 different substituents).

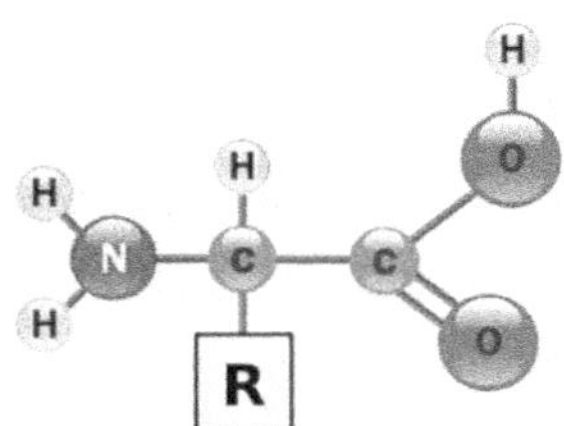

Amino acid structure with the α-carbon attached to the side chain R

D: *cysteine* contains sulfur in the side chain and is the only amino acid forming disulfide bonds (i.e., S–S).

40. A is correct.

During glycolysis, two net (four gross = total) molecules of ATP are produced by *substrate-level phosphorylation*, and two molecules of NAD^+ are reduced (i.e., gain electrons) to form NADH.

B: during glycolysis, *two pyruvates* form from each starting molecule of glucose.

C: *glycolysis* is an anaerobic process and does not require O_2.

D: *glucose* is partially *oxidized* (not reduced) into *two pyruvate* molecules during glycolysis.

E: *pyruvate* is produced during glycolysis (not in the Krebs cycle), converted to acetyl CoA (coenzyme A) that enters the Krebs cycle.

41. B is correct.

Krebs (TCA) cycle and the *electron transport chain* are metabolic processes in the mitochondria.

42. C is correct.

Feedback inhibition occurs when a product binds an enzyme to prevent it from catalyzing further reactions.

Since it is *allosteric inhibition*, this product binds a site other than the active site and changes the conformation of the active site to inhibit the enzyme.

Although the allosteric inhibitor can be the same product of the enzyme-catalyzed reaction *or* the product of the metabolic pathway, it is usually the final product that inhibits the entire metabolic pathway, thus saving the energy necessary to produce the intermediates as well as the final product.

When levels of the final product are low, the enzyme resumes its catalytic activity.

43. A is correct.

Obligate anaerobe reactions occur in the *absence* of O_2.

Oxidative phosphorylation occurs during the electron transport chain and requires a molecule of *oxygen* as the final acceptor of electrons shuttled between the cytochromes.

44. E is correct.

Peptide bonds join adjacent amino acids.

The amino acid (written on the left by convention) contributes an amino group ($\sim NH_2$).

The other amino acid (written on the right by convention) contributes a carboxyl group ($\sim COOH$) that undergoes condensation (i.e., joining of two pieces to form a connected unit) *via* dehydration (i.e., removal of H_2O during bond formation).

45. A is correct.

Oxidative phosphorylation (O_2 as substrate/reactant) occurs only in the electron transport chain (the last step for cellular respiration) in the *inner membrane* (i.e., cristae) of the mitochondria.

NADH and $FADH_2$ donate electrons to a series of cytochrome (i.e., protein) molecules embedded in the inner membrane and create an electron gradient, establishing a proton (H^+) gradient within the *intermembrane space* of the mitochondria.

H^+ gradient drives a proton pump coupled with an enzyme, producing ATP via *oxidative phosphorylation*.

Electrons from NADH and $FADH_2$ are transferred to ½ O_2 (i.e., oxidative phosphorylation) to generate ATP and form H_2O as a metabolic waste (along with CO_2) from cellular respiration.

H_2O and CO_2 (as metabolic wastes) are expired from the lungs during breathing.

Each Krebs cycle $FADH_2$ yields two ATP, while each NADH yields three ATP in the electron transport chain.

NADH from glycolysis only produces two ATP because energy is expended to shuttle the NADH from glycolysis (in the cytoplasm) through the double membrane of the mitochondria.

46. C is correct.

Enzymes do not affect free energy (ΔG); decreasing activation energy increases the reaction rate.

Enzymes do *not* affect the position or direction of equilibrium; they only affect the speed of equilibrium.

47. C is correct.

Zymogen (or proenzyme) is an inactive enzyme precursor. Proteolysis (i.e., cleavage) of the zymogen makes it active to catalyze reactions. Changes in substrate concentration regulate enzyme activity; an increase in substrate concentration increases enzyme activity (at saturation, V_{max} is achieved).

Post-translational modifications occur during protein biosynthesis (after translation), and they may involve cleaving proteins or introducing new functional groups to arrive at the mature protein product.

The enzyme may not be functional until post-translational modifications, a crucial regulatory aspect.

48. B is correct.

$2°$ structure consists of numerous local conformations, such as α-helixes and β-sheets.

Secondary conformations are stabilized by hydrogen bonds or the covalent disulfide bonds between S–S of two cysteines to form cystine.

To be classified as $2°$ structure, the bonds must be formed between amino acids within about 12-15 amino acids along the polypeptide chain.

Otherwise, interactions (e.g., H bonds or disulfide bonds) that connect amino acid residues of more than 15 amino acids along the polypeptide chain qualify as $3°$ protein structure.

If the bonds occur between different polypeptide strands, the interactions qualify as $4°$ protein structures (e.g., two α and two β chains in hemoglobin).

49. E is correct.

Glycogen is a polysaccharide (i.e., carbohydrate) storage form with *highly branched glucose monomers*.

Extensive branching in glycogen provides numerous ends to the molecule to facilitate the rapid hydrolysis (i.e., cleavage and release) of individual glucose monomers when needed *via* epinephrine (i.e., adrenaline).

Glycogen is synthesized in the liver because of high plasma glucose concentrations and stored in muscle cells for release during exercise.

A: *glycogenesis* involves the synthesis of glycogen.

B: *glycogenolysis* involves the degradation of glycogen.

D: plants produce *starch* (i.e., analogous to glycogen) as their storage carbohydrate.

50. B is correct.

The fraction of occupied active sites for an enzyme equals V/V_{max} which $= [S]/([S] + K_m)$.

If $[S] = 2K_m$, the fraction of occupied active sites is 2/3.

51. C is correct.

Allosteric regulators bind enzymes at allosteric sites (i.e., other than active sites) with *non-competitive inhibition*. Once bound to the allosteric site, the enzyme changes its conformational shape.

If the shape change causes the ligand (i.e., molecule destined to bind to the active site) to bind to the active site less efficiently, the modulator is *inhibitory*.

Modulator is excitatory if the shape change causes the ligand to bind to the active site more efficiently (i.e., the active site becomes open and accessible).

52. B is correct.

Biological reactions can be *exergonic* (releasing energy) or *endergonic* (consuming energy).

Biological reactions have *activation energy*, the minimum amount required for the reaction.

These reactions use *enzymes* (i.e., *biological catalysts*) to lower activation energy and increase the reaction rate.

53. C is correct.

Oxidative phosphorylation (in the electron transport chain) uses ½ O_2 as the ultimate electron acceptor.

Electron transport chain (ETC) is on the inner mitochondrial membrane. It uses cytochromes (i.e., proteins) in the inner membranes of the mitochondria to pass electrons released from the oxidation of NADH & $FADH_2$.

Mitochondrial matrix (i.e., the cytoplasm of mitochondria) is the site for the Krebs cycle (TCA).

Outer mitochondrial membrane does not directly participate in oxidative phosphorylation or the Krebs cycle.

Glycolysis and Krebs cycle produce ATP via *substrate-level phosphorylation*, while *oxidative phosphorylation* requires O_2 and occurs during the ETC.

54. D is correct.

Enzymes are often proteins and function as biological catalysts at an optimal temperature (physiological temperature of 36 °C) and pH (7.35 is blood pH).

At higher temperatures, proteins *denature* (i.e., unfolding by disrupting hydrogen and hydrophobic bonds and changing shape) and lose their function.

Proteins often interact with inorganic minerals (i.e., *cofactors*) or organic molecules (i.e., *coenzymes or tightly bound prosthetic groups*) for optimal activity.

Mutations affect *DNA sequences* that encode proteins, resulting in a change in the amino acid sequence of the polypeptide and a change in conformation within the enzyme.

A change in the conformation of enzymes often results in a change in function.

55. D is correct.

Gibbs free energy is:

$$\Delta G = \Delta G° + RT\ln Q$$

where $\Delta G°$ = the Gibbs free energy change per mole of reaction for unmixed reactants and products at standard conditions (i.e., 298K, 100kPa, 1M of each reactant and product), R = the gas constant (8.31 J·mol^{-1}·K^{-1}), T = absolute temperature and Q = [product] / [reactant]

56. D is correct.

Allosteric enzymes have allosteric sites that bind metabolites.

Allosteric sites are distinct from the active sites that bind substrate.

An enzyme must have a *quaternary structure* (multiple polypeptides) to bind more than one molecule.

Allosteric enzymes show *cooperative* substrate binding, with the binding of a molecule at one site affecting the binding at another.

If an enzyme binds an inhibitor at an allosteric site, this *decreases the affinity* of the active site for the substrate.

57. C is correct.

Monoamine oxidases (MAO) are a family of enzymes that catalyze the oxidation of monoamine neurotransmitters. Monoamine neurotransmitters and neuromodulators contain one amino group connected to an aromatic ring by a two-carbon chain ($\sim CH_2–CH_2$).

Monoamines are derived from the aromatic amino acids phenylalanine, tyrosine, tryptophan, and thyroid hormones by aromatic amino acid decarboxylase (enzyme). Monoamines trigger crucial components such as emotion, arousal, and cognition.

Monoamine oxidases are linked to some psychiatric and neurological disorders.

High or low levels of MAOs are associated with schizophrenia, depression, attention deficit disorder, substance abuse, and migraines. MAO degrades serotonin, melatonin, norepinephrine, and epinephrine.

Excessive levels of catecholamines (e.g., epinephrine, norepinephrine, dopamine) leads to hypertensive crisis (e.g., hypertension), and excessive serotonin leads to serotonin syndrome (e.g., increased heart rate, perspiration, dilated pupils).

58. C is correct.

ΔG for hydrolysis of ATP to ADP and Pi is -7.3 kcal/mole.

59. C is correct.

I: *prosthetic groups* are tightly bound chemical compounds required for the enzyme's catalytic activity, and it is often involved at the active site.

II: *zymogens* are inactive enzyme precursors requiring biochemical changes to become active enzymes.

This activation is usually a proteolysis (protein cleaving) reaction that reveals the active site.

60. B is correct.

Hydrolysis of ATP $\rightarrow$ ADP + phosphate forms glucose-6-phosphate from glucose and phosphate.

Enzymatic coupling of reactions allows ATP hydrolysis energy to drive the synthesis of glucose-6-phosphate.

61. A is correct.

Protein function can be altered by *post-translational modifications* (often in Golgi) after synthesis.

The common post-translational modification in eukaryotes is *phosphorylation* which forms the "*mature*" protein. Phosphorylation is the "activator" *or* "deactivator" of proteins/enzymes.

Secondary structure of proteins (e.g., alpha helix and beta pleated sheets) is stabilized by hydrogen bonds and hydrophobic interactions.

Quaternary structure requires two or more polypeptide chains.

Integral membrane proteins are embedded within the plasma membrane. Integral proteins are amphipathic (i.e., hydrophobic and hydrophilic portions). They have a similar structural motif with a hydrophobic domain, α-helical or β-sheet, which traverses the hydrophobic core of the lipid-bilayer membrane.

62. D is correct.

CO_2 is a product of alcohol fermentation (ethanol) and pyruvate oxidation (lactic acid) under aerobic conditions.

In yeast, *alcoholic fermentation* occurs in an anaerobic (i.e., without oxygen) process, producing ATP and the waste products ethanol (alcohol) and carbon dioxide (CO_2).

Pyruvate oxidation occurs in the mitochondria and cytosol in eukaryotes and prokaryotes, respectively.

Pyruvate dehydrogenase complex (PDC) has enzymes that oxidize pyruvate to acetyl-CoA, NADH, and carbon dioxide waste products.

Under anaerobic conditions (e.g., exercise in humans), pyruvate is converted to 2 moles of *lactic acid*, and NADH is oxidized (i.e., loss of hydrogen atom) to NAD^+.

63. E is correct.

Glucose and fructose have different chemical properties despite the same *molecular formula* ($C_6H_{12}O_6$).

Molecular formula of fructose and glucose are identical; however, the *structural formulas* (shown below) are different, giving rise to many *physical and chemical differences* between the two monosaccharides.

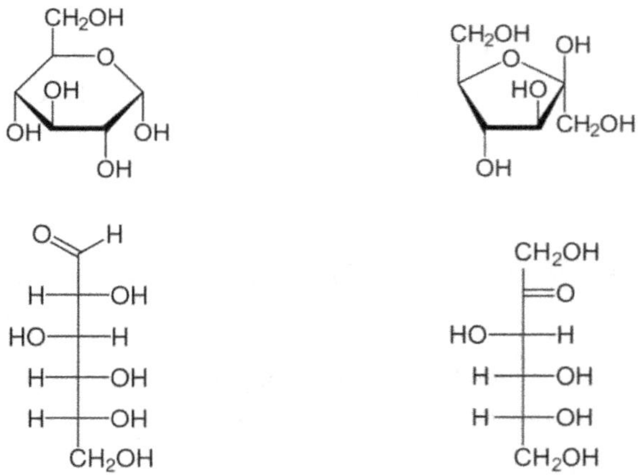

Haworth projections (above) and Fisher projections (below) for glucose on the left and fructose on right

Glucose has an aldehyde (i.e., terminal carbonyl) functional group; fructose is a ketone (i.e., internal carbonyl).

Compared to glucose, fructose has an increased density and decreased melting point with an increased BP.

Glucose, commonly referred to as blood sugar, is the "*preferred*" metabolic energy source.

Insulin is secreted from the *beta cells of the pancreas* in response to increased blood sugar (i.e., glucose) and facilitates glucose's entry into cells. Fructose does *not* stimulate the release of insulin as glucose does.

Glucose fuels cellular respiration, but fructose does not.

Glucose relies on glucokinase or hexokinase enzymes.

Fructose is more lipogenic and relies on fructokinase to initiate metabolism.

Fructose is the sweetest natural carbohydrate.

Notes for active learning

Notes for active learning

Essential Biology Self-Teaching Guides

Eukaryotic Cell & Cellular Metabolism

Molecular Biology & Genetics

Nervous & Endocrine Systems

Circulatory, Respiratory & Immune Systems

Digestive & Excretory Systems

Muscle, Skeletal & Integumentary Systems

Reproduction & Development

Microbiology

Plants & Photosynthesis

Evolution, Classification & Diversity

Ecology & Population Biology

Visit our Amazon store

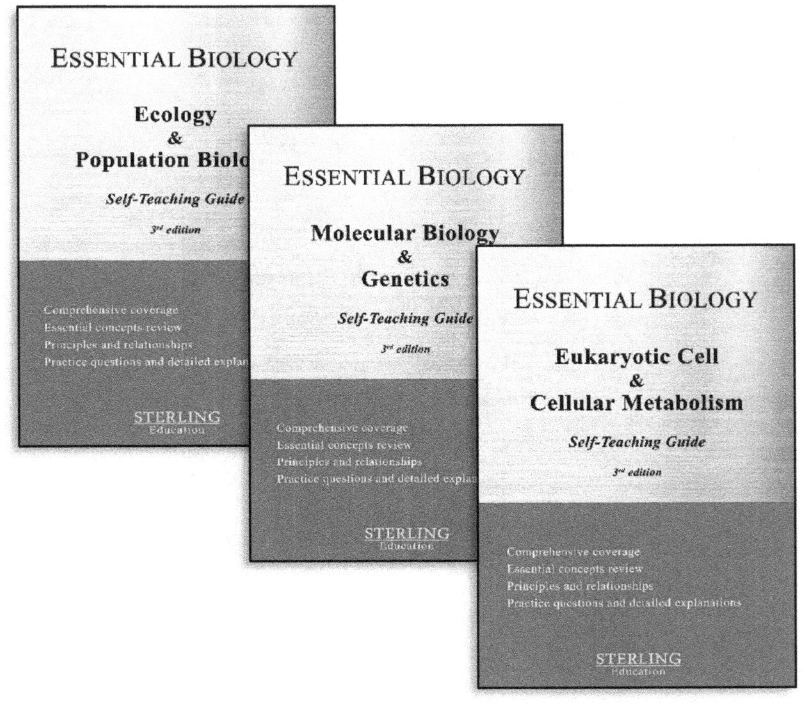

Essential Physics Self-Teaching Guides

Kinematics and Dynamics

Equilibrium and Momentum

Force, Motion, Gravitation

Work and Energy

Fluids and Solids

Waves and Periodic Motion

Light and Optics

Sound

Electrostatics and Electromagnetism

Electric Circuits

Heat and Thermodynamics

Atomic and Nuclear Structure

Visit our Amazon store